Animal KINGDOM

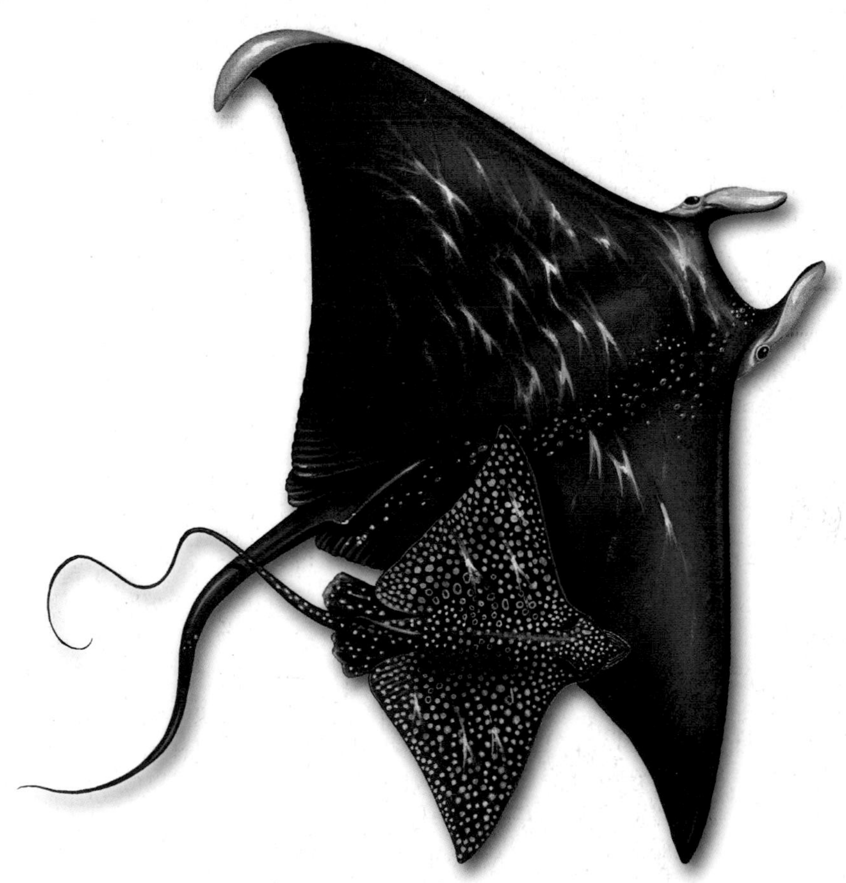

Animal KINGDOM

Steve Parker &
Martin Walters

BARDFIELD
PRESS

First published as hardback in 2000 by Miles Kelly Publishing Ltd

This edition published in 2007 by Bardfield Press

Bardfield Press is an imprint of Miles Kelly Publishing Ltd
Bardfield Centre, Great Bardfield, Essex, CM7 4SL

2 4 6 8 10 9 7 5 3 1

British Library Cataloguing-in-Publication Data
A catalogue record for this book is available from the British Library

ISBN 978-1-84236-941-8

Editorial Director Belinda Gallagher
Art Director Jo Brewer
Editor Steve Parker
Assistant Editor Lucy Dowling
Junior Designer Candice Bekir
Researcher/Indexer Jane Parker
Production Manager Elizabeth Brunwin
Reprographics Anthony Cambray, Liberty Newton, Ian Paulyn

Printed in China

www.mileskelly.net
info@mileskelly.net

The publishers wish to thank Ted Smart for the generous loan of his illustrations.
Illustrators include Andy Beckett, Jim Channell, John Franics, George Fryer, Mick Loates, Robert Morton, Colin Newman, Mike Sauders, Christian Webb.

Contents

Contents

Contents

World of animals

▸ Within the pages of this book all the animals shown in the main picture are listed in this panel. They are named in alphabetical order.

A lonely iceberg floats in the cold polar sea. Suddenly a penguin pops up from the water and hops onto the berg. A seal also hauls itself out onto the ice. A seabird lands for a rest. Just under the waves swim fish, squid and other ocean creatures. Far from being empty of life, the scene is swarming with animals.

Many types of wild places, or habitats, are the same. Deserts by day seem deserted and lifeless. But in the cool night creatures emerge to creep, hop, run, flap, slither and sidewind in search of food and mates. The interior of a tropical forest is gloomy and still, and animals seem absent.

Jaguar
All (or most) of the animals pictured in this book have their own entries, giving important details about their lifestyles, where they live, what they eat and how they breed.

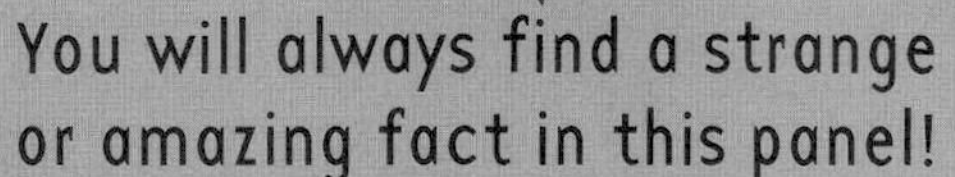

You will always find a strange or amazing fact in this panel!

However they are all around – amazingly camouflaged as dead leaves, bits of moss, creepers and lumps of bark. All of these habitats and many more are covered in this book. They range from the snowy treeless tundra of the far north to steamy tropical swamps, airy broadleaved woodlands, high and rocky mountains, wide grassy plains, and watery places from rivers and lakes to the open ocean.

Each major habitat scene is followed by selected groups of animals, to highlight the incredible range of wildlife and the wonderful ways creatures adapt to their particular surroundings. From snakes to snails, albatrosses to quails, mice to whales – they are all in this book.

Near the North Pole

- Arctic fox
- Arctic hare (mountain or blue hare)
- Arctic tern
- Grey wolf
- Harp seal
- Musk ox
- Norway lemming
- Polar bear
- Ptarmigan
- Reindeer (caribou)
- Snow goose
- Stoat
- Walrus

Animals of ice and snow

Tundra lands border the Arctic Ocean, mainly in North America and Asia.

The top of the world is a large, shallow and very cold sea, the Arctic Ocean. It is covered with a vast floating raft of ice which is, on average, 5–7 m thick. The lands around the ocean are the far northern parts of Asia, North America and Europe. These vast, treeless, boggy places are called tundra. They enjoy a brief warm summer when the ice raft shrinks to cover the central part of the ocean, at the North Pole itself. But as autumn approaches the ice raft spreads and a big freeze grips the tundra lands. Many larger animals travel or migrate south away from the frost and snow. Those who stay endure some of the harshest conditions on Earth.

Arctic tern

This bird could also be called the Antarctic tern. It breeds in the short Arctic summer, then flies halfway around the world to spend another summer in Antarctica, resting and feeding. As autumn begins, off it goes again north to the Arctic. No other creature migrates so far every year. (See also page 26.)

Snow goose

Snow geese are some of the earliest spring arrivals in northern North America, to lay their eggs as the thaw gathers speed. They are stocky, sturdy birds with a beak-tail length of about 70 cm. As the summer fades they fly south to spend winter in slightly warmer coastal regions. (See also page 18.)

Harp seal

These seals live in the North Atlantic and Arctic Ocean and hardly ever come out onto land. They rest and even breed on floating pack ice or icebergs. They eat mainly fish and also shellfish such as crabs and shrimps, diving more than 100 m deep for several minutes. Harp seals grow to 1.8 m long. (See also page 206.)

Polar bear

The near-white coat of this bear provides excellent camouflage in Arctic ice and snow. Most bears have a varied diet of meat and also some plant foods. The polar bear is the most carnivorous or meat-eating – simply because there are very few large plants in its watery habitat. In the wild it feeds mostly on seals and fish. However some bears have learned to scavenge around towns and rubbish dumps. An adult male polar bear has a head-body length of 3 m and weighs more than 400 kg. Females are slightly smaller. The mother polar bear digs a cave in a snow bank where she gives birth to her 1–2 cubs in mid winter. She stays in the snow den with them for 2 months, living off her stores of body fat. (See also page 20.)

Arctic hare

The Arctic hare is known by other names, such as tundra hare and mountain hare, depending on where it lives. In fact another of its names is varying hare, also blue hare due to the blue-grey sheen of its fur in summer. This hare nibbles at any plant food and is active mainly at night. (See also page 192.)

Some Arctic terns fly a total of 30,000 km each year, from the Arctic to the Antarctic and back again. They are the most-travelled creatures in the world.

Musk ox

With its long and shaggy outer coat, the musk ox is well protected from the bitingly cold tundra winds. It also has another layer of shorter, thicker fur beneath, the undercoat. Musk oxen live in small herds in northern North America. Adults are about 2 m head-body length.

Stoat

The long-bodied, short-legged stoat is an extremely fierce hunter of rabbits, hares, lemmings and similar mammals, and birds as big as ptarmigan. It moults its brown fur in autumn to become white in winter, when it's known as the ermine.

Grey wolf

The wolf's eerie howl carries through the cold, still Arctic night. Wolves venture out onto the tundra in summer after prey such as hares, young musk oxen, reindeer and birds. The wolf pack trots south to the shelter of the great pine forests for winter. (See also page 196.)

Ptarmigan

Like many birds around the Arctic, the ptarmigan sheds or moults its feathers twice each year. In spring it changes to its summer plumage of mottled brown to match the bushes and plants. In autumn it grows its mainly white winter feathers to blend in with the snow.

Norway lemming

This small creature is a rodent – a close relative of mice and voles. It lives in Northern Europe and eats seeds, grasses, mosses and other plant food. In winter lemmings dig tunnels under the snow so they can keep eating.

Arctic fox

Unlike most larger land animals of the Arctic, which seek shelter in woods to the south, the Arctic fox can stay out on the tundra all winter. But to survive it must eat a variety of foods, from lemmings and small birds to carcasses of reindeer, seals and stranded whales.

Reindeer

This deer is usually known as the caribou in North America, where the main wild herds roam. Herds are also kept by people in many northern lands, especially Europe and Asia, to provide milk, meat, furs and skins. Reindeer are the only deer where females, as well as males, have antlers.

Walrus

The walrus is the largest member of the seal and sea-lion family, called pinnipeds. It lives along Arctic coasts and swims with its flipper-like limbs. A very thick layer of fat, called blubber, under the skin keeps in its body heat as the walrus plunges into the cold sea.

The tusks of the walrus are very long upper teeth called canines. They can reach 60 cm in length, although the male walrus is much larger than the female and so he has longer, thicker tusks. Walruses live in groups or herds of up to about one hundred. They often haul themselves out onto rocky shores or icebergs, to bask in the weak Arctic sun. (See also page 16.)

DO LEMMINGS COMMIT MASS SUICIDE?

There are tales of huge numbers of lemmings throwing themselves off cliffs or jumping into rivers, as though they want to die. But they are really following their natural urges and searching for a new place to live. This is because every few years when conditions are good, lemmings breed very fast. They eat all the food and the area gets too crowded with them. Some lemmings set off to find more food and space. Even cliffs and rivers do not daunt them.

Stocky seabirds with paddle wings

- Atlantic puffin
- Black guillemot
- Brunnich's guillemot (thick-billed murre)
- Crested auklet
- Guillemot (common murre)
- Horned puffin
- Little auk (dovekie)
- Razorbill
- Tufted puffin

Auks are plump, stocky, upright, waddling seabirds – the northern versions of the penguins which live in the far south, in Antarctica. Unlike penguins, auks have not lost the power of flight. But their wings are only just large enough to keep them in the air, and they have to flap very fast to stay aloft. This is because an auk's wings are also adapted as paddles for swimming and diving, to hunt for fish beneath the waves. The squat body and thick layers of tight-packed feathers keep in body warmth and keep out the cold waters of the northern seas. Auks breed on seaside cliffs and ledges where most predators cannot reach their eggs or chicks.

Atlantic puffin
The puffin's deep, narrow beak has very sharp edges to hold several fish at once. The beak's colour is much brighter in spring to attract a breeding mate.

Little auk
These starling-sized seabirds breed in their millions on rocky slopes around Greenland, Iceland and Svalbard. They feed on tiny animals like shrimps.

Brunnich's guillemot
This auk stays in the cold waters of the far north even in winter. Its main plumage is so dark brown that from a distance it looks black.

Horned puffin
The horned puffin has an even more massive bill than its Atlantic cousin. In summer it breeds along the coasts of Alaska, then it flies as far south as California for a warmer winter.

Razorbill
Razorbills are found around the coasts of the North Atlantic Ocean. The sharp-edged and sharply-hooked beak is well named since it can easily slash or peck through human skin.

Guillemot
Unlike Brunnich's guillemot, this bird flies south in winter from the coldest Arctic seas. It dives under the surface to catch fish, worms and shellfish. Its single egg is laid on a bare cliff ledge.

Black guillemot
This auk stays close to shore, where it dives deep after fish. In winter its plumage changes from mainly black to largely white and it almost looks like a gull.

Crested auklet
This auk lives on the Aleutian Islands in the North Pacific between Alaska and Russia. The curved crest is larger in summer to impress a mate during courtship.

Tufted puffin
The huge red and yellow bill looks even more extraordinary with the yellow head feathers curving back from behind each eye.

The biggest auk, known as the great auk, could not fly. It was killed by sailors for its meat, eggs and feathers. Great auks died out for ever in 1844.

Flipper-footed fish-eaters

- Leopard seal
- Northern elephant seal (male and female)
- Steller sea-lion (male, female and young)
- Walrus (male and female)

Seals, sea-lions and the walrus belong to a mammal group known as pinnipeds, which means 'flipper-feet'. They are slow and awkward on land but in water they swim with tremendous speed and grace. Sea-lions 'row' with their front flippers. On land they can prop themselves up on these flippers and waddle along. Seals are even better swimmers and use their rear flippers, but they can only wriggle or 'hump' on land. Most pinnipeds live along coasts and hunt fish, squid and similar prey.

Steller sea-lion
This is the largest type of sea-lion, the male growing to almost 3 m in length. It lives along the shores of the North Pacific Ocean and dives down nearly 200 m in search of squid, octopus and fish. As in many sea-lions and seals, the males come ashore at the start of the breeding season and fight each other to gain a patch of beach, called a territory. Males without territories have little chance of mating with females.

Leopard seal
With its leopard-like spots and wide, sharp-toothed mouth, this Southern Ocean seal is 3.5 m long. It's a fierce, fast hunter, but not of fish. It preys on penguins, seabirds – and other seals.

Walrus
Like many pinnipeds, the male walrus is much larger than the female. He grows to 3.3 m long and over 1200 kg in weight. A walrus uses its tusks to lever shellfish from sea-bed rocks. It also eats crabs, starfish, worms and other bottom-dwelling creatures. Walruses live all around the Arctic Ocean.

Northern elephant seal
The male elephant seal with his huge, floppy nose grows to a massive 5 m in length. Most of his weight, as in other pinnipeds, is a thick layer of fatty blubber under the skin. This keeps in body warmth when swimming in cold water. It also makes the body shape more streamlined for faster swimming.

Elephant seals show the greatest size difference between the sexes of almost any animal. Males weigh more than 2300 kg. Females are less than one-third this size.

Honking, walking and pecking

- Barnacle goose
- Bar-headed goose
- Bean goose
- Brent (brant) goose
- Canada goose
- Egyptian goose
- Emperor goose
- Greylag goose
- Pink-footed goose
- Red-breasted goose
- Snow goose
- White-fronted goose

Geese are between swans and ducks in size, and they spend much time walking about and feeding on the ground. In spring they fly north to breed on the icy, treeless plains called the tundra that surround the Arctic Ocean. During the long days of summer food is plentiful here. But the summer only lasts a few weeks, so in early autumn the geese fly south back to warmer regions. These long seasonal journeys are called migrations. For the winter the geese live near lakes, rivers, estuaries and coasts. They often form large, honking flocks as they waddle steadily across marshes and fields, pecking for food. Sometimes they damage farm crops.

Canada goose
As its name suggests, the Canada goose came originally from North America. But it has spread to Britain and northern Europe, where it is familiar as a semi-tame bird in parks, gardens and farmyards. It is one of the largest geese, and its long neck gives it a swan-like appearance in flight.

Red-breasted goose
This very colourful goose breeds in Siberia. It builds its nest and rears its goslings near to the nest of a bird of prey such as a rough-legged buzzard. The buzzard might attack the chicks. But the risk is worth it, since the buzzard also keeps away Arctic foxes who are more likely to eat the goslings.

Emperor goose
These small, stocky geese breed around the coasts of the Bering Sea, where the far east of Russia is within sight of Alaska in North America. They like to nest on coastal marshes and tundra. In the winter the emperor goose sometimes wanders as far south as the warm coasts of California.

Brent goose
Known as the brant goose in North America, this is a small and active member of the group with dark plumage. It prefers coastal areas with mud flats and salt marshes rather than inland regions, and it sits and bobs on the sea like a duck. The Brent goose especially likes to eat seaweeds and eelgrass.

Barnacle goose
Barnacle geese are very noisy, making barks and yaps that sound like small dogs fighting. They breed mainly in Greenland and Svalbard (Spitsbergen). Before people discovered such remote nesting areas, they thought these geese hatched from the barnacles that grow on boats and driftwood.

Snow goose
Snow geese are well named from their almost pure white feathers. A flock of thousands of these beautiful birds can look like a snowstorm. Snow geese nest in the Arctic in the far north of North America, laying eggs even before the snow melts. They migrate to the coasts farther south for the winter.

Geese are sometimes kept as 'guard birds'. They honk loudly, flap their wings and peck anyone who seems suspicious. They also honk at cars and other vehicles.

Far from cute and cuddly

- American black bear
- Asian black bear (Himalayan black bear)
- Brown bear (includes grizzly, Kodiak bear, Kamchatka bear, Alaskan bear)
- Sun bear (Malayan sun bear)
- Polar bear
- Sloth bear
- Spectacled bear

Bears may look amusing and cuddly, especially as 'teddy-bear' toys. But real bears are big, strong, powerful and sometimes very dangerous. They are best left alone. There are seven kinds of bears and they are all very similar. They have large heads, dog-like faces, big jaws, sharp teeth, bulky bodies, sturdy legs and very strong claws. Bears find meals mainly by smell. They like meat or fish but they also eat many other foods including fruits, berries and the honey of wild bees. Apart from the polar bear of the icy north, bears live in woods and forests. They are shy and rarely seen in the wild. They have also been hunted by people so most types are now rare.

Asian black bear

The Asian black bear has a jet-black coat of soft, silky fur with white markings on its chest. It also has larger, more rounded ears than other bears. It is an agile climber and often clambers into the branches to rest. This bear eats mainly fruit, nuts, shoots, grubs and insects, also birds and their eggs.

American black bear

This bear is found in North America, from Canada south to Mexico. Although some black bears really are black, others are brown, rusty-red, grey-blue or even creamy white. They like to stay in forests but they sometimes venture out to scavenge on dead farm animals or waste dumps.

Brown bear

Brown bears are found in woods and forests in Asia, North America and Europe. The largest types are the grizzly of North America and the Kodiak bear of Kodiak Island near Alaska. Europe's brown bears are smaller and now very rare, found only in a few isolated hills and mountain regions.

Polar bear

This bear's yellow-white fur blends with the snow and ice of its Arctic home. It helps the bear to creep up on its prey, mostly seals. It also hunts fish and swims well, the thick fur and blubber (fatty layer under the skin) keeping it warm in the freezing water. Some polar bears hang around towns and feed on scraps.

Sun bear

The smallest bear, at about 100 cm long, the sun bear of Southeast Asia lives in tropical forests. It feeds on shoots, fruits, birds, mice and other small animals. It also licks up termites.

Sloth bear

Eastern India and Sri Lanka are home to the unusual sloth bear with its long, shaggy fur. It often hangs upside down like a sloth using its long, curved claws.

Spectacled bear

The only bear from the southern half of the world, in South America, this type has pale markings around its eyes. It lives mainly in mountain forests.

The brown bear and polar bear are the largest meat-eating animals on land, even bigger than the Siberian tiger. A polar bear may be 3 m long and weigh 800 kg.

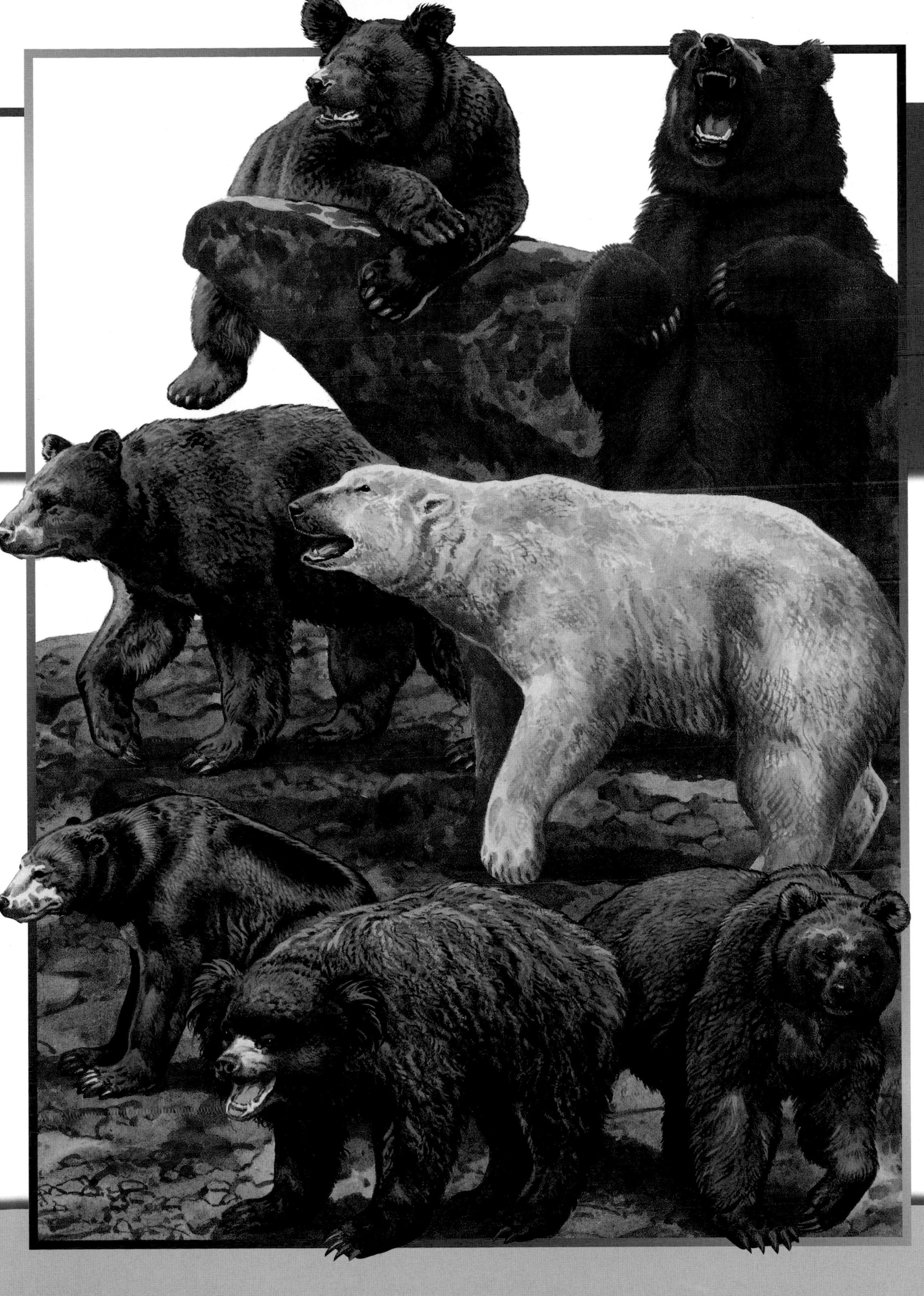

At the water's edge

- Avocet
- Black-tailed godwit
- Common sandpiper
- Curlew
- Lapwing (peewit)
- Oystercatcher
- Redshank
- Stone curlew

Waders are truly shore birds, specialized for feeding at the water's edge. Most live along the mud flats, salt marshes, beaches and rocky coasts of the sea but some are found near lakes, ponds and marshes inland. A typical wader has long, thin legs to walk in shallow water without getting its feathers wet, and a long, thin beak to jab and probe in soft mud or sand for worms, shellfish and other small creatures. Some waders have shorter beaks and peck rapidly for tiny creatures on the surface. Many waders breed in the Arctic during the brief summer, then travel or migrate south to spend the winter along the shores of milder regions.

Redshank
One of Europe's commonest waders, the redshank really does have bright orange-red legs – and a beak to match. It nests on high, boggy moors and coasts.

Oystercatcher
The oystercatcher uses its strong, chisel-shaped bill to hammer into mussels and oysters. It levers open the shell parts to get at the soft flesh inside. It can also crack open crabs and sea snails.

Stone curlew
The stone curlew is an unusual wader in two ways. It is found in dry, sandy or stony places like heaths and moors. It is also active at twilight and night.

Black-tailed godwit
This large wader breeds on marshes in Europe and Asia. Its long bill can probe deep into damp soil to pull out worms and grubs.

Avocet
The avocet's beak has a unusual upward curve. The bird swishes it from side to side while holding it at the surface or just under the water, to filter out its food of tiny animals.

Lapwing
The lapwing looks black and white at a distance, but has beautiful glossy green and purple upper parts. It often walks across farmland pecking for soil animals.

Common sandpiper
An active bird of rivers and lakes, the common sandpiper stands on a waterside boulder and bobs its head up and down as it feeds. It is a small and very shy wader. If disturbed it flits away quickly, low across the water's surface, making its high-pitched 'twee-wee-wee' alarm call.

Curlew
One of the biggest waders, the curlew uses its long, down-curved bill to extract grubs and worms from deep in soil, sand and mud. Curlews often gather on the mud flats of an estuary (river mouth). As they wheel across the sky they make their flight call which is a sad, lonely-sounding, rising whistle: 'coor-lee'.

The oystercatcher's beak is so strong that it can break open most kinds of shelled animals, including crabs and snails. It can even knock limpets off rocks.

Noisy whales of northern seas

- Beluga (belukha or white whale) (female and calf)
- Narwhal (tusked whale) (male)

The beluga and the narwhal make up a small group called white whales – although the narwhal is mottled cream, blue, grey or brown on its back and sides. They are both unusual among whales because they have no dorsal (back) fin, and they can bend their necks to look around and curl their lips into facial expressions. Also they live in very cold water among the icebergs and pack ice of the Arctic Ocean, especially along shores and estuaries and in shallower coastal waters. The beluga in particular is a very noisy whale. It makes a huge variety of squeaks, moos, clicks, chirrups and bell-like clangs that can even be heard above the water.

Narwhal

The narwhal has only two teeth. In the male one of these, the upper left incisor, keeps growing and becomes a long, sharp tusk with a twisted spiral pattern. It can reach nearly 3 m in length. The other tooth stays about 20 cm long. The whale's body may grow to more than 5 m long and weigh 1.5 tonnes. The female narwhal also has two teeth and one may grow into a tusk, but this is much shorter and may not even grow beyond her lips. A narwhal uses its lips and tongue for feeding, not its teeth.

Like belugas, narwhals live in small groups of adults and young. These often gather into larger herds of hundreds. The herds migrate along the coast with the seasons, to follow the shoals of fish and other food and to keep clear of the spreading winter ice.

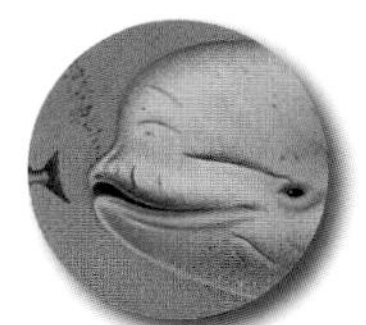

Beluga

The beluga's many sounds and noises help it to keep in touch with other members of its group or herd. The sounds also bounce off nearby objects like rocks and animals and the beluga hears the echoes. Like a bat in air, this echolocation system helps the whale to find its way in dark or muddy water and to detect food. The beluga also 'makes faces' at other herd members by opening and twisting its mouth into a variety of smiles, frowns and grins. Belugas grow to 5–6 m in length and eat a wide range of food including fish, squid, worms, shellfish, shrimps and crabs. They feed mainly on the bottom of the sea and use their pursed lips to suck worms from the mud and shellfish out of their shells.

Beluga calf

A newborn baby beluga is 80–90 cm long. It is also brown in colour, which changes to slatey blue-grey by about one year old. Then it gradually lightens to pure white by its adult age of five years. The youngster feeds on its mother's milk for up to two years and hardly ever leaves her side during this time.

UNICORN OF THE SEA

The narwhal's long tusk looks like the head horn of the mythical horse known as the unicorn. Exactly why this whale has a tusk is not clear. Sometimes males come to the surface and use their tusks like swords to 'fence' each other. This may be to gain success in mating with females.

About one male narwhal in 50 has two tusks.
It looks like a combination of whale and walrus!

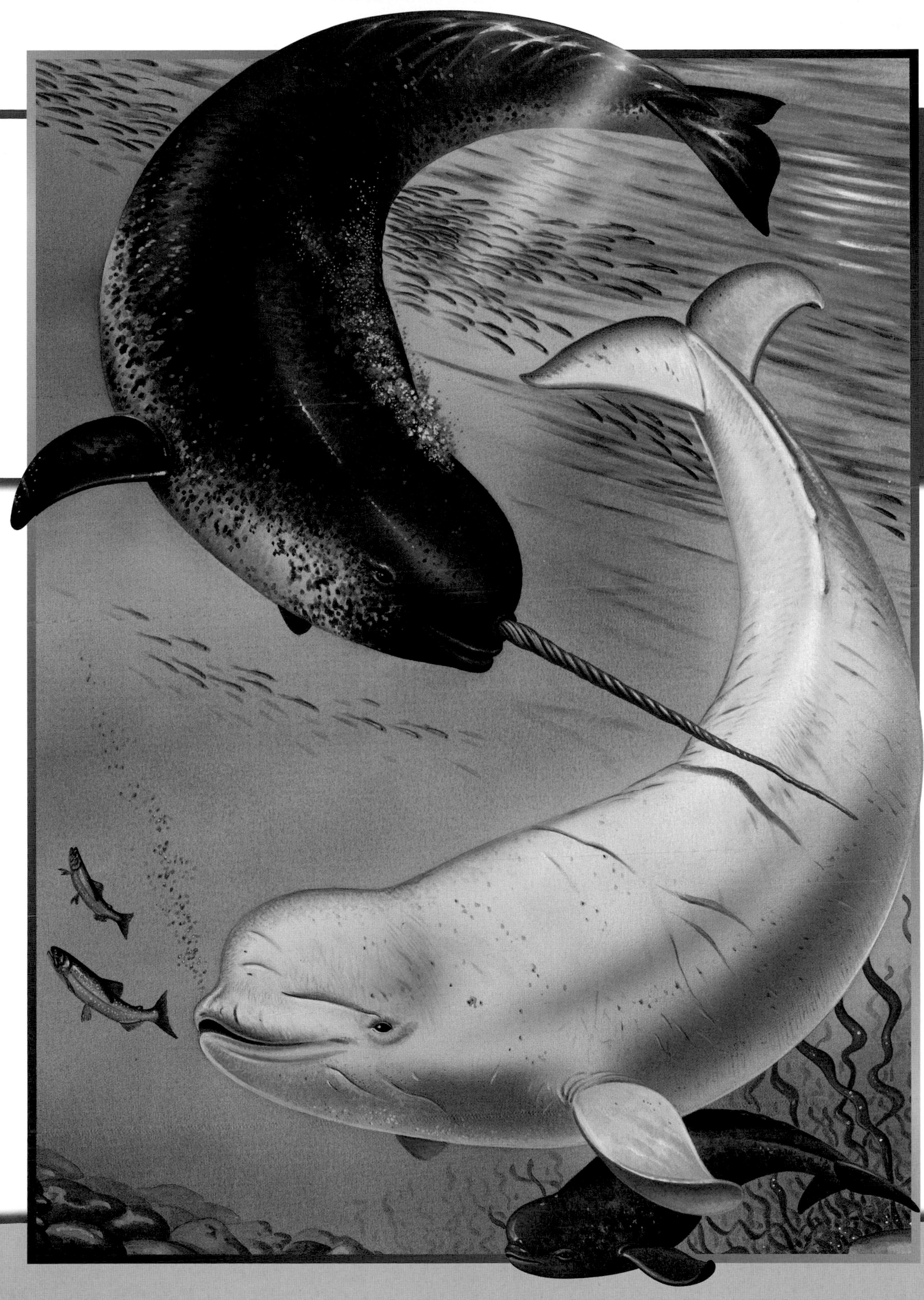

Over shores and seas

- Arctic skua
- Arctic tern
- Black-headed gull
- Black tern
- Common tern
- Great black-backed gull
- Herring gull
- Long-tailed skua

Gulls are the most familiar birds at the seaside – and often the noisiest! They also hunt for grubs and worms on fields inland, especially in the winter, and some kinds scavenge in rubbish for scraps. Most gulls have white and grey plumage, and some have black heads or backs. The larger types have powerful hooked beaks to catch smaller birds and other animals. Skuas are similar to large gulls. In North America they are called jaegers, or 'hunters', because they chase other birds to make them drop their food, which the skua then eats. Terns are small, slim, graceful seabirds. Their forked tails give them the nickname of 'sea-swallows'.

Herring gull
This is Europe's commonest large gull. It breeds in fields and on dunes, shingle banks, rocky ledges and cliffs. In towns it even nests on buildings – and never sees the sea!

Long-tailed skua
This skua is slim and delicate, with narrower wings than other skuas. It is very elegant in flight and migrates north to breed on the Arctic tundra.

Common tern
This bird is very similar to the Arctic tern apart from the black tip to its red beak. It breeds in colonies on sandy coasts, islands, and sometimes on dunes.

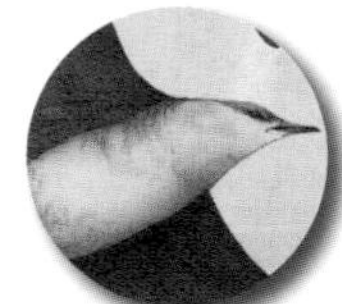

Arctic skua
The commonest European skua, this type nests in the far north on open tundra and moors. Many skuas have long central tail feathers which may help their twisting flight movements.

Black-headed gull
The 'black' head of this gull is really dark chocolate brown in the breeding season. In winter it often visits parks and gardens.

Black tern
Like a large dark swallow, the black terns hunt over lakes and marshes. It hovers and skims the surface, dipping its beak for small water creatures.

Arctic tern
A bright red beak and long tail streamers (feathers) identify this small seabird. It has a dainty appearance but it makes very harsh, grating calls. This tern breeds in the Arctic and north, and as far south as Britain and Ireland. In the autumn it migrates to the Southern Ocean and some birds even reach Antarctica.

Great black-backed gull
One of the largest gulls, this strong bird has a black back and wings. With its hooked beak it can feed on a wide range of food, including carrion such as dead seals or whales washed up on the shore. It breeds on rocky and stony coasts, particularly on small and remote islands.

The sooty tern of tropical seas may remain in the air for five years, sleeping and feeding on the wing. It just lands briefly to breed.

Trees and leaves

- Bank vole
- Bluethroat
- Cat snake
- Common centipede
- Common toad
- Fallow deer
- Great black slug
- Grey squirrel
- Hermann's tortoise
- Mistle thrush
- Pipistrelle bat
- Privet hawkmoth
- Purple emperor butterfly
- Pygmy shrew
- Roman snail
- Stoat
- Weasel
- Wild cat
- Wood mouse
- Wryneck

Woodland creatures

Temperate woods once covered much of Europe, until replaced by farms.

After the medium-long, medium-cold winter of a temperate climate, spring comes to the deciduous wood. 'Deciduous' means trees lose their leaves in the cold season. As days grow warmer their buds burst and the scene brightens with new leaves, flowers and blossoms. Animals who have spent most of the winter asleep in nests and burrows, or scratching through the frost and snow for food, become active again. Also various birds return after their winter migration to warmer regions such as Africa. Late spring and summer are a time of feeding and breeding. Then the chilly autumn winds blow dead leaves from the trees and the winter shutdown returns.

Wild cat

Like a large and heavy pet cat, the wild cat stalks the forest at night in search of mice, rats, small birds and similar prey. With a head-body length of 60 cm it is a powerful hunter, snarling and hissing at enemies. Wild cats live in most of Europe and also in Africa and across the Middle East to India.

Fallow deer

The male fallow deer or buck has broad, spreading antlers. The female, or doe, lacks antlers as in most other types of deer. Fallows are common in woods, especially in country parks and in forests conserved for wildlife. They came originally from Western Asia but they have been introduced into many areas.

Bluethroat

A shy bird, the bluethroat skulks in bushes and undergrowth and even stays low as it flies across clearings. It pecks at leaves and soil for grubs, bugs and worms, and also eats berries in autumn. Bluethroats live in scattered parts of Western Europe, and from Eastern Europe across Asia and into Alaska.

Grey squirrel

Although they are now common in Britain and a few places in Europe, grey squirrels were only brought there in the late 1800s from North America. They are larger and stronger than the local red squirrels, and also better suited to living in deciduous woods. Red squirrels are now found mainly in pine woods.

Pipistrelle bat

The pipistrelle is Europe's smallest bat, with wings only 20–22 cm from tip to tip. It is also one of the most common and adaptable bats, found in woods, scrub, heath and along the banks of rivers and lakes. It has taken to living in buildings too, even in towns where it darts around street lights to hunt small insects.

Purple emperor

A majestic butterfly, the purple emperor is usually seen flitting along the tops of oak trees in the summer sunshine. It is probably a male chasing away rivals and flashing his shiny wings to attract a female. She has brown rather than purple wings but her white patterns and eye spots are the same.

Wild cats look very similar to large pet tabby cats. But the original ancestor to pet cats, more than 5000 years ago, was probably the African (Abyssinian) wild cat.

Wryneck

The odd name for this bird comes from its habit of twisting, bobbing and drooping its neck as it looks for danger or food, or to attract a mate. It perches on a tree trunk and leans back on its tail like a woodpecker, as it flicks insects from bark with its long tongue.

Hermann's tortoise

Tortoises may be slow and lumbering, but when they pull their head and legs into the strong shell, few predators can break in. Hermann's tortoise has a shell about 20 cm long and lives mainly in dry woods around the Mediterranean region.

Weasel

A mini-version of the stoat, the weasel is small and bendy enough to dash down a mouse hole after its prey. Weasels are often helpful to the farmer because they keep down the numbers of animals which might eat crops, such as mice, voles, rats and baby rabbits.

Privet hawkmoth

Hawkmoths are named after their large swept-back wings and fast, direct flight (like an insect version of a real hawk). The adult moth rests on any plant but the large, green, pink-striped caterpillars feed mainly on privet leaves.

Cat snake

The cat snake has a poisonous bite but this is not very harmful to people. In any case like most snakes it usually slithers away if disturbed. Cat snakes grow to about 90 cm long, live in dry, rocky places in Southeast Europe and eat mainly lizards.

Common toad

Toads visit ponds and lakes in spring to breed. For the rest of the year they live on land, usually in moist places like damp meadows, hedges and low-lying woodlands. But they can survive in dry areas too such as heath and scrub. They eat all kinds of small animals, from slugs, worms and flies to baby mice and bird chicks.

Common centipede

There are many kinds of common or lithobius centipedes in woodlands around the world. They scuttle through the dead leaves and soil at night as they hunt slugs, worms and grubs. By day they hide in damp places under stones and bark.

Mistle thrush

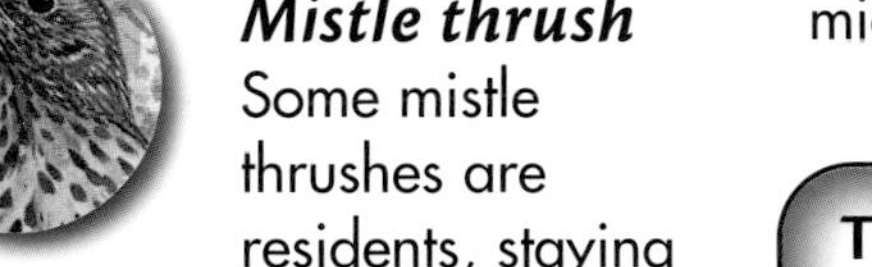

Some mistle thrushes are residents, staying in the same place through the year. Others are migrants, flying north in summer to breed in woodland, then returning south to warmer places for the winter. These birds sing powerful, flute-like notes from the treetops.

THE DAWN CHORUS

A deciduous woodland is one of the finest places to hear the dawn chorus of bird song, especially in late spring. Birds sing to attract mates for breeding, and also to tell nearby birds to keep out of their particular patch of land, or territory. In a European wood as the sky lightens the first birds to sing are usually blackbirds, then song thrushes and redstarts, followed by wrens and robins.

Wood mouse

As the sun sets and the diurnal (day-active) creatures hide in their nests and holes, nocturnal animals begin their nightly search for food. Wood mice are plentiful in most woodlands. They eat seeds, berries, nuts, leaves and other plant parts and also take small animals such as slugs, snails and woodlice.

The name 'centi-pede' means '100 legs'. But there are hundreds of kinds of centipedes and their legs vary in number from about 30 to more than 350.

Squawking thieves

- Azure-winged magpie
- Blue jay
- Common jay
- Common magpie
- Nutcracker
- Plush-crested jay
- Turquoise jay

The crow family (see page 36) contains many large perching birds, including jays, magpies and nutcrackers. Many of the jays have bright, gaudy plumage and are easy to spot by sight – and by sound. But they are not famous for their beautiful songs. They make loud, hoarse, grating calls which sound like a person coughing with a sore throat! The birds use these harsh calls to keep contact with each other in woodland and scrub. Magpies and jays are adaptable eaters and take a very wide range of food, from seeds and nuts to insects, frogs and mice. They also steal the eggs and babies of other birds, which makes them pests in some regions.

Azure-winged magpie
This magpie is found in China and Japan – and, oddly, also halfway round the world in Portugal and Spain. It may have been brought back to Europe by exploring sailors returning from the Far East. Azure-winged magpies are very social, which means they form flocks with others of their kind.

Common jay
Jays live and breed mainly in mixed broadleaved woods across Europe and Asia. They have also spread into wooded parks and gardens, where they swoop between trees with a flash of their blue wing fronts and harsh calls. Like squirrels, jays often gather acorns and other nuts and bury them to eat later.

Turquoise jay
The Americas have more than 30 kinds of jays, mainly in the tropics. The turquoise jay is one of several threatened by loss of their forest home.

Blue jay
Common in woods and gardens across eastern North America, this jay feeds mainly on seeds and nuts. But it also raids birds' nests for eggs and chicks.

Nutcracker
Nutcrackers live in mixed woodland, from the uplands of Scandinavia across to the hills of eastern Europe. In autumn they gather piles of tree nuts and seeds, especially arolla pine.

Common magpie
Shiny black-and-white plumage and a long, green, glossy, blue-tipped tail are a magpie's main features. It prefers open country with trees and hedges and avoids thick woods. The female and male build a large twig nest high in a tree or bush. This has a loose, dome-shaped roof and a side entrance.

Plush-crested jay
The soft black feathers on the head of this jay form a crest which feels like a velvet cushion! Plush-crested jays live in forests in many parts of South America. Like several jays and magpies they copy the calls of other birds in their loud 'babble', and even imitate other animal sounds.

The plush-crested jay can mimic many sounds including frog croaks, monkey screeches and even (like parrots and mynahs) the spoken words of the human voice.

Shy shapes in the woods

- Asian mouse deer (larger mouse deer or greater chevrotain)
- Water chevrotain
- Chinese water deer
- Moose (European elk)
- Muntjac
- Red deer (wapiti or American elk)
- Roe deer

Slim, shy and speedy leaf-eaters, with tall antlers, long legs and brownish dappled coats – this describes most of the 43 types of deer around the world. They are mainly woodland creatures which live in small family groups or herds, in all habitats from snowy tundra to tropical forests. The antlers of deer differ from the horns of cows and antelopes because they do not grow continuously. They are shed one year and grow again the next. Also most antlers are branched. Nearly all male deer have them except for the four types of tiny mouse deer, or chevrotains, and the three musk deer. No female deer have antlers except for the reindeer, also called the caribou.

Moose
This is the largest deer, known as the elk in Europe. (The elk in America is a different type of deer, shown on the right.) Unlike many other deer, the moose tends to live alone. It dwells in the depths of conifer woods during the winter, eating buds, twigs and mosses. In summer it feasts on water plants.

Chinese water deer
This deer is one of the few where males lack antlers. But both sexes have tusks, which are extra-large canine teeth. Water deer usually live in female-male pairs near lakes or rivers, feeding on reeds, rushes and other water plants. The female has up to four babies rather than one as in other deer.

Red deer
The red deer lives in all northern lands and also in Africa, with different names such as wapiti, maral or Bactrian deer. Herds feed mainly on leaves, buds, shoots and other soft plant parts. In the breeding season the stags (males) roar and bellow at each other, clash antlers and fight to mate with the females.

Muntjac
Standing about 40–50 cm tall at the shoulder, the muntjac is found in India and Southeast Asia. It was also brought to Britain, mainly to dwell in game parks, but it has since escaped to live wild. It feeds in the late evening and early morning on low-growing bushes and grasses, and rests by day.

Roe deer
Most deer have short tails which they usually keep pressed against their back legs. The roe deer has almost no tail at all. It lives across Europe, Southern and Southeast Asia.

Asian mouse deer
Mouse deer are more similar to camels and pigs than to real deer. They hide in thick forest and eat mainly flowers and fruits.

Water chevrotain
This African chevrotain is hardly larger than a hare, standing about 30 cm tall at the shoulder and only 10 kg in weight.

A well-grown male moose can stand more than 2 m tall at the shoulder and weigh over 600 kg. The female is about three-quarters this size.

The cleverest birds?

- Chough
- Hooded crow
- Jackdaw
- Raven
- Rook

Crows, rooks and ravens are found mainly in open country, including mountains, cliffs, moors and hillsides. Like their cousins the jays and magpies (see page 32) they are large, strong, adaptable birds with powerful beaks and legs, as much at home on the ground as in the air. Ravens are the largest members in the world of the huge group of birds known as perching birds (passerines). A raven is easily powerful enough to kill prey such as rabbits. Ravens and many crows tend to live on their own or with a breeding partner. Others like rooks, jackdaws and choughs lead more social lives. They roost and breed in colonies and feed together in flocks.

Raven

These massive corvids (members of the crow family) reach 65 cm in length. They can kill prey but usually scavenge on dead animals. In olden times they pecked at the bodies of criminals who had been hanged, so the raven became known as a bird of ill omen. Its deep, hollow call carries far over its upland home.

Rook

Rooks are similar to carrion crows except for a patch of pale skin at the beak base and feathered, shaggy-looking legs. Rooks also live mainly in groups whereas a crow tends to be alone. Rooks like open country with scattered fields and woods where they can find their main food of grubs, slugs and worms.

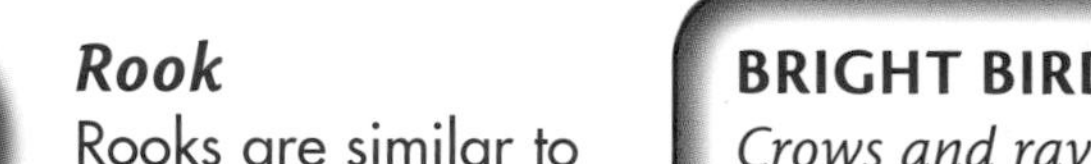

BRIGHT BIRD!

Crows and ravens use many tricks to gather food. Crows fly high with shellfish and drop them on rocks to crack them open. They wait near gull breeding colonies and dash in to steal unguarded chicks. Tests have shown that ravens can count up to five or six, outscoring birds like parrots.

Hooded crow

The hooded crow has a similar lifestyle to the all-black carrion crow but lives mainly in north, east and south-east Europe, also in north-west Scotland and Ireland. It prefers open hills and moors where it eats many foods from seeds to snails. Hooded and carrion crows may breed together.

Jackdaw

Jackdaws are noisy and active. They like to roost and breed in tree holes and on ledges along cliffs and rocky outcrops. Window ledges and chimney pots resemble these wild sites, which is why jackdaws have spread into towns and cities. They are also famous for 'stealing' shiny objects like coins and rings.

Chough

Choughs, with their distinctive red legs and red beaks, live in loose flocks. They are skilled, acrobatic fliers and utter high-pitched squealing calls. They live in rocky mountainous areas and along remote coasts in Europe. Choughs have become rare due to loss of the natural meadows where they feed on soil animals.

A flock of ravens is kept at the historic Tower of London in England. Legend says that if the ravens fly away, England will be taken over by invaders.

Small, busy insect-eaters

- European hedgehog
- European mole
- Hispaniolan solenodon
- Pygmy shrew (least shrew)
- Pyrenean desman
- Rufous elephant shrew

Shrews, hedgehogs, moles and similar animals are known as insectivores. This name means 'insect-eaters' but many of these small, busy, active, darting mammals feed on a variety of tiny prey including worms, snails, slugs and spiders. There are about 345 kinds (species) of insectivores around the world. They include tenrecs, moonrats and desmans – as well as over 200 types of shrews, from giants to pygmies! Most insect-eaters have long, pointed, quivering, whiskery noses, little eyes and ears, and very sharp teeth. They are mainly active at night and they use their keen senses of smell and touch, rather than sight, to catch their prey.

Pygmy shrew

Pygmy shrews are the smallest of all land mammals, weighing only 2 g and measuring hardly 6 cm long – including the tail! They are fierce hunters and feed on worms five times their size, as well as insects. They are so small and active, and use up so much energy, that they must feed every 4–6 hours – or starve!

Hispaniolan solenodon

Solenodons look like rat-sized giant shrews. There are only two kinds, found in the Caribbean, and both are very rare. They use their very flexible snouts to sniff out insects from cracks and crevices. They can also pounce on larger prey like lizards and mice using their long, sharp claws.

Rufous elephant shrew

Elephant shrews were thought to be shrews, but they are so unusual that they are now put in their own mammal group. They run swiftly to escape predators along special pathways which they keep clear by regularly removing twigs and leaves. Their large eyes give them keen vision even in dark forests.

European hedgehog

These familiar garden predators feed mainly on worms, slugs and caterpillars. They are usually helpful to the gardener but they may be poisoned by eating slugs which have fed on slug pellets. Rolling into a tight, prickly ball protects the hedgehog from predators like foxes and stoats.

Pyrenean desman

The desman is like a water-living mole with a longer nose and a flattened tail. It has waterproof fur and webbed feet and swims very fast after small fish, tadpoles, pond snails and water insects. There are only two kinds of desman, one in the Pyrenees between France and Spain and the other in Asia.

European mole

A mole is seldom seen unless floods drive it to the surface. It spends most of its life in its large burrow network, feeding on worms, grubs and other soil creatures. The mole's front feet are almost like shovels to push aside earth. Molehills are loose soil that the mole thrusts up from its tunnels – which stretch up to 150 m.

The prickles of a hedgehog are specially thick, hardened, pointed hairs. A fully grown hedgehog has about 16,000 of them.

Plump birds of parks and cities

- Bar-shouldered dove (mangrove dove or river pigeon)
- Crested pigeon
- Galapagos dove
- Laughing dove
- Wood pigeon (ring dove)
- Olive pigeon
- Racing pigeon
- Town pigeon (domestic pigeon)
- Pied imperial pigeon (nutmeg pigeon)

Pigeons and doves are familiar garden birds and also well known from the tame pigeons which flock in our town parks and squares. There are 300 different kinds (species) in the group, ranging from large crowned pigeons in New Guinea to tiny ground doves almost as small as sparrows. The larger, plumper types are usually called pigeons and the smaller, slimmer species are doves. They are mainly seed-eaters and build untidy tree nests of sticks and twigs. Most pigeons and doves have soothing but slightly sad-sounding songs. They are very strong fliers with amazing direction-finding instincts and are bred to race each other and even carry messages.

Wood pigeon
Found across Europe, this large pigeon has white patches on its wings and neck which show up clearly in flight. It lives in woods and also in most other habitats, from upland moors to coastal sand dunes. Wood pigeons often gather in large flocks outside the breeding season, searching for seeds, shoots and other plant food in fields and farms. Some come into city parks and are almost as tame as town pigeons.

Racing pigeon
The racing pigeon is a form of the town pigeon, bred over many generations for its speed and also for its remarkable homing ability to find its way back to its cage or 'loft'.

Pied imperial pigeon
Also known as the Torres Strait pigeon, this bird lives in the area of that name, in northern Australia. It feeds mainly on fruits.

Olive pigeon
This large, dark pigeon is found in the forests and woodlands of East Africa. It likes to roost and nest in thorn trees and also raids wheat fields.

Bar-shouldered dove
A long-tailed dove of coasts along north and east Australia, this dove prefers scrubby creeks and mangrove swamps.

Town pigeon
Town pigeons are descended from wild rock doves which live naturally on cliffs and other rocky sites. But walls and window ledges do just as well!

Laughing dove
This is one of the smallest doves, found across much of East Africa. It lives in scrub and woods and visits gardens and farmland. It is also found around the southern shores of the Mediterranean Sea and as far north as Turkey. Pigeons and doves are unusual among birds in that they produce a kind of 'milk' inside the crop (the bag-like part of the lower throat) to feed their young chicks.

Racing pigeons can fly at steady speeds of more than 75 km/h and cover great distances – 10,000 km or more.

Woodpeckers would peck wood

- Acorn woodpecker
- Black-headed woodpecker
- Gilded flicker
- Greater yellow-naped woodpecker
- Orange-backed woodpecker
- Red-collared woodpecker
- Red-throated wryneck
- White-backed woodpecker
- White-bellied sapsucker

Most woodpeckers live up to their name and peck at trees with their chisel-shaped beaks, to find grubs and other small animals in bark and wood. They also chip out nesting holes in trunks. These activities help many other woodland birds, which come to feed under the loose bark or nest in old woodpecker holes. A woodpecker's foot has two toes pointing forwards and two backwards for an extra-secure grip when holding onto an upright tree trunk. Also the tail feathers are stiff and pointed so the woodpecker can lean back on them when climbing or hammering. And the bones and muscles in its head and neck are very strong, so it doesn't get a headache!

Acorn woodpecker This woodpecker chips holes in trees and pushes acorns and other nuts into them, to eat when food is scarce in winter.

Red-collared woodpecker The forests of south China and much of Southeast Asia are the haunts of this woodpecker. Its red neck collar is very distinctive.

White-backed woodpecker A woodland bird from north-east Europe and Russia, this type likes forests with plenty of old, rotting trees and lakes or streams.

Orange-backed woodpecker Like several woodpeckers this bird is often found by the loud tapping sounds it makes. It can pull or strip back large sections of bark to look for grubs underneath.

Gilded flicker Flickers are large, common North American birds. They are bold and often seen in parks and gardens as well as in woodlands. The gilded flicker's underwings have a golden sheen.

White-bellied sapsucker The sapsucker drills rows of holes in a tree, then licks up the sugary sap which oozes out using its brush-like tongue. It also gobbles up insects which feed on the sap.

Red-throated wryneck Wrynecks resemble woodpeckers but lack stiff tail feathers. They twist their heads oddly to look around when alarmed, hence their name.

Greater yellow-naped woodpecker Despite its colourful feathers, this woodpecker is hard to spot when it stands still in its forest home of eastern Asia.

Black-headed woodpecker This noisy bird has a loud, squawking call in flight, when its bright red rump contrasts with the green body and red-capped black head.

A single green woodpecker may eat more than 2000 ants in one day.

Small, bright, busy and noisy

Titmice (tits) are common garden birds with bright plumage and busy, acrobatic behaviour. They often hang under and clamber over bird-feeders and breed in garden nest-boxes. In America some types are known as chickadees. Bulbuls, another group of active songbirds, are larger than tits with bubbling, tuneful songs. They live mostly in Africa and southern Asia. Wrens are tiny, shy and skulk in thick bushes where their brown plumage makes them hard to see. But you can certainly hear them – for their size, wrens are the loudest of all birds.

Red-whiskered bulbul
Originally from India and Southeast Asia, this bulbul has been brought to new areas around Sydney, Australia and Miami, USA. It is tame and often comes to gardens.

Blue tit
Found in Europe, North Africa and Asia, this tit is one of the most common visitors at the bird table. Despite its small size it often chases much larger birds away from the food.

European wren
One of Europe's smallest birds, the wren is found in many habitats from open moor to dense marsh. It holds its tail almost upright and builds a domed nest among tree roots.

Long-tailed tit
Small and delicate, this tit builds an amazing melon-shaped nest from spiders' webs, moss and animal fur. The nest is lined with soft feathers.

Common iora
Found in India and Southeast Asia, the common iora is familiar in gardens and also lives in woods and mangrove swamps. It has a clear, whistling call.

Crested tit
The crested tit lives mainly in conifer woods, especially among pine or spruce trees. It eats the seeds from the cones as well as snapping up insects.

Golden-fronted leafbird
Named after its orange forehead, this bird hunts insects in the forest trees of South and Southeast Asia. It often copies other birds' calls.

Black-capped chickadee
One of the best-known garden birds in North America, this tit is named after its black head and 'chickadee' call.

Fairy bluebird
These bluebirds like to eat tropical fruits, especially figs. They are always on the move, hopping and flitting through forests in Southern Asia.

The tiny long-tailed tit gathers more than 2000 soft feathers to make its beautifully woven nest.

An Eastern forest

- Asian purple emperor butterfly
- Azure-winged magpie
- Black-naped oriole
- Black-veined white butterfly
- Blue-and-white flycatcher
- Brahmaea certhia moth
- Broad-billed roller
- Brown hairstreak butterfly
- Collared scops owl
- Comma butterfly
- Daurian redstart
- Dusky thrush
- Eurasian flying squirrel
- Grey-backed thrush
- Grey-faced buzzard
- Japanese pygmy woodpecker
- Limenitis moltrechti butterfly
- Mugimaki flycatcher
- Pale-legged warbler
- Papilio macilentus butterfly
- Racoon-dog
- Red-throated flycatcher
- Siberian blue robin
- Sika deer
- Sparrowhawk
- Tiger
- Tiger shrike
- Tristram's bunting
- White-throated rock-thrush
- Wild boar
- Yellow-billed grosbeak
- Yellow-rumped flycatcher
- Yellow-throated marten

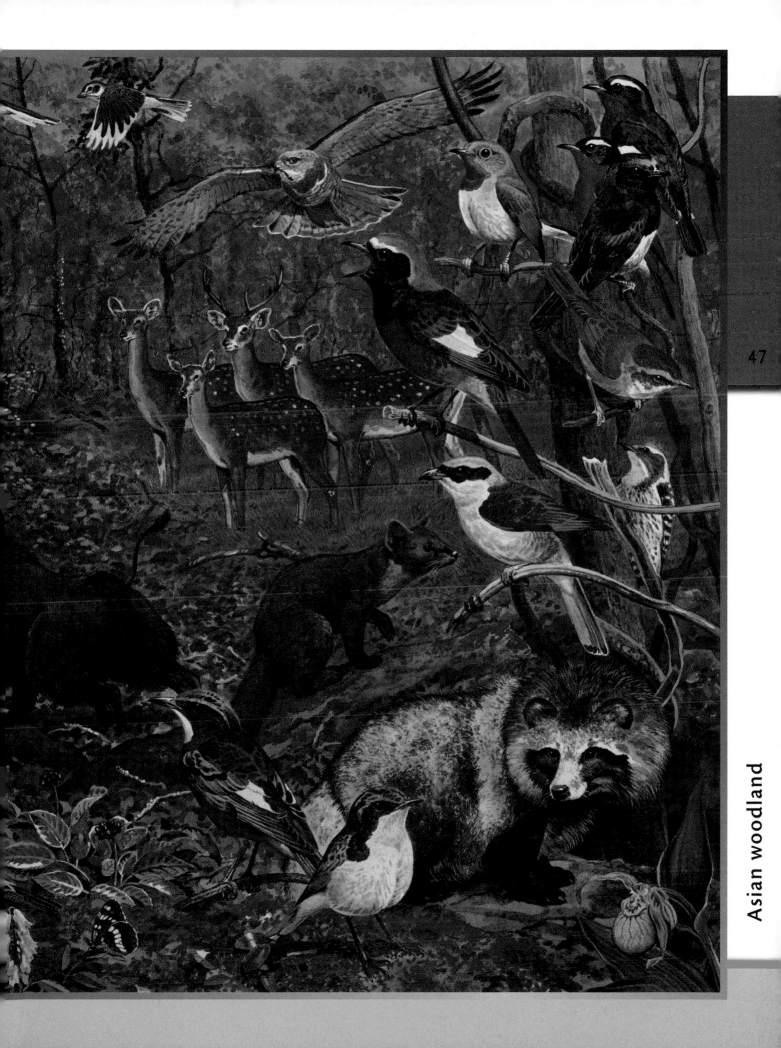

Among the branches

Temperate woods in the far east of Asia are hemmed in by grassland and desert.

Among the trees, something stirs. Is it a tiger or a tiger shrike? Both are in their own ways just as deadly. The mid summer leaves of the Asian woodland conceal many surprises. Deer browse peacefully and butterflies flit across sunlit clearings. But in nature, the struggle for survival never ends. A flycatcher grabs a butterfly, then almost at once a sparrowhawk swoops on the flycatcher. The battle is against predators like the yellow-throated marten, buzzard and raccoon-dog. But half a year later when the leaves and shelter have gone, the main enemy will be the elements – sharp frosts, deep snow, icy winds and long, dark, freezing nights.

Blue-and-white flycatcher
The intensely coloured feathers of this flycatcher, in different shades of blue with a white chest, are designed to attract a mate at breeding time rather than for camouflage. Even so the flycatcher is difficult to see since it lurks in the undergrowth, only dashing out for a few seconds to pursue food.

Sika deer
Also known as Japanese deer, sikas live across East and Southeast Asia. They have been taken to new regions such as New Zealand. They live in family groups, the chief male identified by his large antlers. He stands about 100 cm tall at the shoulder. In winter the deers' coats become greyer all over.

Brahmaea certhia moth
The brahmaea moths are mostly large, strong fliers with wingspans of 15 cm or more. This type has a delicate grey-brown pattern to match the bark of the trees, where it usually rests during the day with its wings held out sideways. It comes to flowers and tree blossom to feed on the nectar.

Tiger
Top predator of East Asian woods, this is the largest variety of tiger and the biggest of big cats – the Siberian tiger. Very few survive in the most remote regions. Its main prey include wild boar, deer and wild cattle. The tiger hunts mainly at twilight or in darkness, but also during the day in winter. (See also page 146.)

Azure-winged magpie
About 35 cm from beak to tail, this bird is a typical member of the magpie group. It chatters and calls loudly as it hops about the wood, pecking for insects, worms, nuts, berries, fruits and various other foods. Oddly, though, the male does the main job of feeding the chicks. (See also page 32.)

Racoon-dog
Adaptable as the raccoons of North America, raccoon-dogs resemble foxes but are really a type of wild dog related to the wolf. They sleep in family groups by day in a hole or den and emerge at night to eat small animals (especially fish and frogs), fruits and berries. They also scavenge in our rubbish.

In autumn a flying squirrel may hide or hoard more than 200 nuts every day, in addition to the food it actually eats.

THE NIGHT SHIFT

As the sun sets and woodland shadows deepen, day-active (diurnal) animals settle to sleep. Their places are taken by an equivalent set of night-active (nocturnal) creatures. Owls take over from sparrowhawks, buzzards and other birds of prey. Nightjars hunt moths, replacing flycatchers who have chased butterflies. A dangerous time is the twilight of dusk or dawn. Eyes take time to adjust to changing light levels. This is when the tiger prowls.

Collared scops owl

This is one of the most widespread of owls, with a huge range across East and South Asia. It catches mainly large insects such as beetles, crickets, grasshoppers, moths and cockroaches. But it also dives onto mice or lizards on the woodland floor, and it can grab a small bat in mid air.

Dusky and grey-backed thrushes

The dusky (above) and grey-backed thrushes flick aside dead leaves to reveal slugs, woodlice and similar tasty items. Like other thrushes they have a springy, bouncy hop.

Papilio macilentus butterfly

There are more than 200 kinds of papilio or swallowtail butterflies. The long, pointed ends of the rear wings resemble a swallow's tail. Red spots warn this one tastes foul.

Yellow-throated marten

Martens are long and supple predators in the mustelid group, related to stoats and weasels. The yellow-throated marten has a head-body length of 60 cm and a tail of 40 cm. It climbs well and eats many kinds of small animals like lizards, mice, voles, birds and their eggs.

Sparrowhawk

This fast, agile flier bursts from its hiding place among the leaves and zigzags through the trees after a sparrow or similar small bird. It may also chase after large flying insects such as beetles or grasshoppers. Sparrowhawks have spread in some areas since they can also adapt to planted conifer forests.

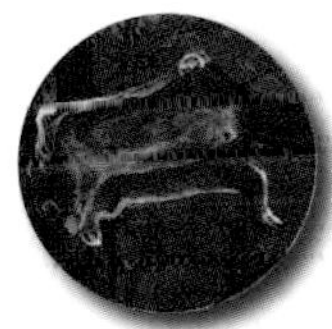

Flying squirrel

The flying squirrels are gliders rather than true fliers. They have huge eyes to see their flight path as they swoop from tree to tree in dusk or darkness, on the outstretched flaps of skin along the sides of the body. Like most squirrels they eat tree parts such as buds, shoots, nuts and soft bark. (See also page 56.)

Wild boar

Few animals fight harder when cornered than the wild boar. Even the tiger is cautious. This stocky, muscular pig grows to about 1.2 m head-body length. It can bite and kick very hard, slash with its tusks and run at speed to trample the enemy. (See also page 164.)

Comma butterfly

The comma spreads its bright orange-brown wings to bask in the sun or to attract a mate. Otherwise it folds its ragged, wavy-edged wings over its back to show their dark grey-brown undersides. This makes the comma look just like a dead leaf.

Red-throated flycatcher

Flycatchers do catch flies, also butterflies and other winged insects. They swoop over leaves and twigs as well, to snatch caterpillars, beetles and bugs. The red-throated flycatcher sings loud and tuneful songs, like its thrush and warbler cousins.

Flying squirrels can glide more than 50 m in 5 seconds, and suddenly fold up their 'wings' and drop into the soft leaves on the forest floor if a hawk appears.

Catching flies in mid air

- Black phoebe
- Buff-breasted flycatcher
- Eastern kingbird
- Eastern wood peewee
- Fork-tailed flycatcher
- Great crested flycatcher
- Great kiskadee
- Least flycatcher
- Scissor-tailed flycatcher
- Superb lyrebird
- Thick billed kingbird
- Vermilion flycatcher

The tyrant flycatchers are American birds, different from the flycatchers of Europe and Asia. With more than 400 kinds or species, they make up the largest family of birds in the Americas. Various tyrant flycatcher species are found from the conifer forests of northern Canada, down through North and Central America to the tip of South America. But the richest diversity is in the tropical rainforests of the Amazon region. As their name suggests, many of these birds chase and catch flies. Some of the larger types prey on creatures as large as crickets and lizards. A few tyrant flycatchers are brightly coloured but most species are fairly dull shades of brown or green.

Buff-breasted flycatcher
This rare bird lives in the dry, rocky canyons of Mexico and southern Arizona. It is very similar to the least flycatcher (shown just to its right in the illustration opposite). However the least flycatcher is much more common and lives in broadleaved woodlands across much of North America.

Black phoebe
The unusual combination of jet-black head, breast and upper body with a white belly area make this small, neat bird easy to identify. It is a common type of tyrant flycatcher in the dry woods of south-western North America. Like its relatives it has a fast, fluttering flight as it pursues small airborne insects.

Great crested flycatcher
With its crest and yellow belly, this is a common and distinctive bird of mixed woodlands, parks and gardens in eastern North America. In a wood it selects an old woodpecker hole or the natural cavity in a hollow tree for its nest site. It has also taken to nest boxes in gardens.

Superb lyrebird
Lyrebirds eat numerous kinds of foods. The superb lyrebird of south-east Australia shows off its amazing tail feathers during its courtship display, and makes a continuous stream of rich, tuneful sounds. It also copies the calls of local birds and even unnatural sounds like car sirens and chainsaws.

Vermilion flycatcher
The male vermilion flycatcher is named after his bright red plumage – red being an unusual colour in the flycatcher group. This colouring is mainly to attract a mate at breeding time. Female and young vermilion flycatchers are drab grey-brown which is much more suitable for camouflage.

Great kiskadee
This noisy bird is one of the largest of the group. Its thick, stout, all-purpose beak shows that it eats a wide range of foods, compared to the thin, delicate beaks of its cousins. As well as flies and other insects the great kiskadee hunts small mammals like mice, also frogs and lizards. It even dives for fish.

The eastern kingbird is one of the most aggressive birds. It readily attacks other birds that enter its territory, even much larger types such as crows and hawks.

Slinky killers

- American badger
- Chinese ferret-badger
- Eurasian badger
- Eurasian otter
- Ferret-polecat
- Little grison (Patagonian ferret)
- Marbled polecat (marbled ferret)
- Ratel (honey badger)
- Sea otter
- Speckle-necked otter (spot-necked otter)

Stoats, weasels, ferrets, polecats and martens are long-bodied, short-legged, sharp-toothed hunters. They are active, flexible and fast-moving and they often race into holes or burrows after prey. Many are extremely fierce, quite ready to attack animals larger than themselves. Otters and mink also belong to this mammal group, which is known as the mustelids. They swim well and prey on fish, crayfish and similar water creatures. Badgers and skunks are mustelids too. They catch small animals, especially earthworms and beetles. But badgers especially have a more varied range of foods and eat fruits, berries and other plant parts.

Ratel
The ratel is also known as the honey badger because it often feeds on the honey of wild bees – and their grubs too. It lives from Africa across the Middle East to India.

Eurasian badger
Badgers are very strong, powerful animals which come out at night. They live in family groups in a huge network of underground tunnels and chambers called a sett.

Sea otter
Sea otters rarely come onto land, and rarely go into water more than 15 m deep. They live along the coasts of the North and West Pacific and eat shellfish, worms, starfish and sea urchins.

Speckle-necked otter
This is probably the fastest and most skilful swimmer of all freshwater otters. It lives in rivers, lakes and swamps in most parts of Africa.

Chinese ferret-badger
Found across Southeast and East Asia, this animal can bite hard and produce a terrible smell when alarmed. It hunts mainly in trees.

Eurasian otter
The otter uses its webbed feet and strong, thick tail to swim speedily after food. It eats mainly fish but also hunts frogs, water voles and small waterbirds.

Little grison
This quick, darting predator is smaller than its cousin the grison from northern South America. The little grison lives in rocky mountains farther south.

American badger
Similar to its Eurasian cousin, the American badger has a dark cheek patch. It eats mainly rats, mice, voles, birds and birds' eggs.

Marbled polecat
The marbled polecat is small for a mustelid, with a head and body only 34 cm long. It lives across Central Asia and hunts mice, voles and lemmings.

The biggest otter is the giant or Brazilian otter of South American rivers and swamps. It grows to 2 m in total length and 30 kg in weight.

Flying death on fast wings

- Barred forest falcon
- Eleonora's falcon
- Laughing falcon
- Red-headed falcon
- Red-legged (collared) falconet
- Saker falcon

Falcons are fast-flying birds of prey. They use speed to chase and catch victims, twisting and turning with amazing aerobatic skill. Large falcons hunt mainly other birds, up to the size of pigeons and partridges. Some of these big and powerful falcons, like the peregrine and saker, are almost the equal of eagles. They swoop and thud their talons into the victim in mid air, in a blur of feathers and blood. Smaller falcons tend to prey on little songbirds and also on large flying insects such as dragonflies, butterflies and beetles. Kestrels hunt by hovering into the wind, head perfectly still and staring at the ground, before diving to pounce on their prey.

Red-legged falconet
This dainty bird is common in open forests that cover the foothills of the Himalaya Mountains and stretch across to Southeast Asia. Falconets are the smallest birds of prey. The red-legged falconet is only 18 cm from beak tip to tail end. Its legs are not red but the feathers just above are.

Red-headed falcon
The striking chestnut-red cap of this bird contrasts with the mainly grey plumage and barred chest as seen in other members of the group. The red-headed falcon is commonest in parts of East Africa. It often nests in palm trees and it feeds mainly on other birds, which it tears up for its chicks.

Barred forest falcon
Found from Mexico through Central America to Argentina, this falcon feeds mainly on lizards and snakes, small birds and mammals such as rats and mice. It has a long tail and short, rounded wings so it can swerve and turn at speed among the close trees of its dense woodland home.

Eleonora's falcon
This long-winged falcon specializes in hunting small birds. It breeds unusually late in the summer, in colonies on rocky islands in the Mediterranean Sea. This late breeding enables the falcon to catch many of the small songbirds flying past, heading south for winter, to feed its own chicks.

Laughing falcon
This falcon's loud, repeated two-part call sounds slightly like the 'ha-ha!' of human laughter. The rainforests of Central and South America are its home, where it catches mainly snakes – including poisonous ones. The falcon's armoured talons (clawed toes) strike too fast for the snake to bite back.

Saker falcon
One of the largest and most powerful falcons, the saker is a bird of open country, especially dry scrub and steppe (grassland). It is found from eastern Europe across Central Asia to China. This majestic hunter seizes ground-living mammals like hares, marmots and pikas as well as other birds.

The peregrine falcon is the world's fastest-moving animal, reaching more than 250 km/h in its power-dive called the 'stoop'.

Sharp claws and bushy tails

- African bush squirrel
- Asiatic striped palm squirrel
- Beaver
- Bush-tailed ground squirrel
- Common marmot
- Eurasian red squirrel
- Malabar giant squirrel (Indian giant squirrel)
- Northern flying squirrel

Not all squirrels are bright-eyed, bushy-tailed and live in trees. But most are. The squirrel family has nearly 270 members and is part of the vast rodent group of mammals. As well as typical tree-dwelling squirrels it includes the bigger, heavier marmots, and also ground squirrels such as prairie dogs and chipmunks which burrow in soil. However most squirrels are amazing climbers, clinging to the bark with their sharp claws and then making huge leaps, using their furry tails for balance and steering in mid air. Flying squirrels have taken this a stage further and swoop between trees using parachute-like flaps of stretched skin along the sides of the body.

Common marmot Marmots are like big ground squirrels, thick-set and thick-furred with short tails. They live in high meadows and rocky scrub.

Eurasian red squirrel With its bright fur and neat ear tufts, the red squirrel is very distinctive. It sometimes buries seeds and nuts, and sniffs them out later in the year from deep in the soil.

Northern flying squirrel Flying squirrels live mainly in dense forests. Unlike most squirrels they tend to be active at night, seeing in the dark with their large eyes.

African bush squirrel This slim, grey, rather rat-like squirrel lives in the Congo region of West and Central Africa. It is at home in trees or on the ground.

Asiatic striped palm squirrel These squirrels are found mainly in Sri Lanka and India. They resemble chipmunks and are always active, searching the forest edge for buds, nuts and seeds.

Bush-tailed ground squirrel Ground squirrels and prairie dogs live in huge networks of burrows and tunnels. They disappear at once into these if danger appears.

Beaver Beavers are not squirrels but close rodent relations. They grow up to 1.5 m long including the scaly tail, which is slapped on the water to warn other family members of danger. Beavers swim well with their webbed back feet. They fell trees by gnawing, for food and to build a dam which makes a home pool.

Malabar giant squirrel This is a very bulky squirrel, weighing up to 3 kg. It often leans forwards from a branch to feed, back feet gripping firmly and balanced by its tail. This leaves its front feet free to grab food. Many squirrels hold items in their front paws as they eat, turning nuts to be cracked open with the teeth.

One red squirrel may strip open more than 150 pine cones in a single day, to get at the pine nuts (seeds) between the cone's flap-like scales.

Scampering through the trees

- Coati (coati mundi)
- Giant panda
- Kinkajou
- Olingo
- Raccoon
- Red panda (lesser or red tree panda)

Raccoons, coatis, kinkajous and ringtails are agile, active creatures similar to dogs or small bears – except they live in trees. They are from the Americas, have long bodies and long tails and are active at night. Raccoons, especially, eat a huge variety of foods from small animals like mice, birds, frogs and fish to shoots, fruits and berries. Most have brown or grey fur, often with mask-like face markings and a ringed tail. Pandas are close cousins of raccoons from Asia. The red panda lives a similar lifestyle to the raccoon. However the giant panda eats bamboo and almost nothing else. It is very, very rare indeed – a world symbol of nature conservation.

Raccoon
Raccoons resemble burglar-masked, long-furred, bushy-tailed mini-bears. Like foxes they are well known as scavengers and often raid trash cans and rubbish heaps for leftover food scraps, usually at night. The raccoon holds its food delicately in its front paws for nibbling and may even wash it before eating.

Coati
Unlike most members of the raccoon group, coatis are usually active by day. They live in the forests of Central and South America. The coati uses its long, flexible nose to sniff out its food of insects and other small creatures from among leaves, rocks and bark. It also likes ripe fruits.

Olingo
The olingo's head and body are about 40 cm long – and so is its bushy tail, used for balancing among the branches. Olingos eat mostly fruits but they also take insects, birds and small mammals like mice. They gather with each other and with kinkajous for feeding trips and rarely come down to the ground.

Giant panda
There are fewer than 1000 giant pandas left in the wild, almost all in western China. They live alone in the thickest bamboo forests, eating the leaves and fleshy stems of these tall grasses. The panda has a 'sixth finger', which is really an extra-long wrist bone, to split open the bamboo shoots.

Red panda
Like a tree-living fox, the red panda has very soft reddish-brown fur. It is found in China, like its giant cousin, and also in nearby countries. It feeds on the ground on bamboo and other grasses, leaves, fruits and nuts, and it also catches small creatures. By day it sleeps in a tree, curled inside its bushy tail.

Kinkajou
One of the strangest members of the raccoon group, the kinkajou of Central and South America has short fur, large eyes and a prehensile or grasping tail. It can climb through trees almost as well as a monkey as it searches for fruits, the sweet nectar inside flowers and the honey of wild bees.

Bamboo is so poor in nutrients that the giant panda spends up to 16 hours each day eating. However if it comes across a dead animal, it helps itself to the rotting flesh.

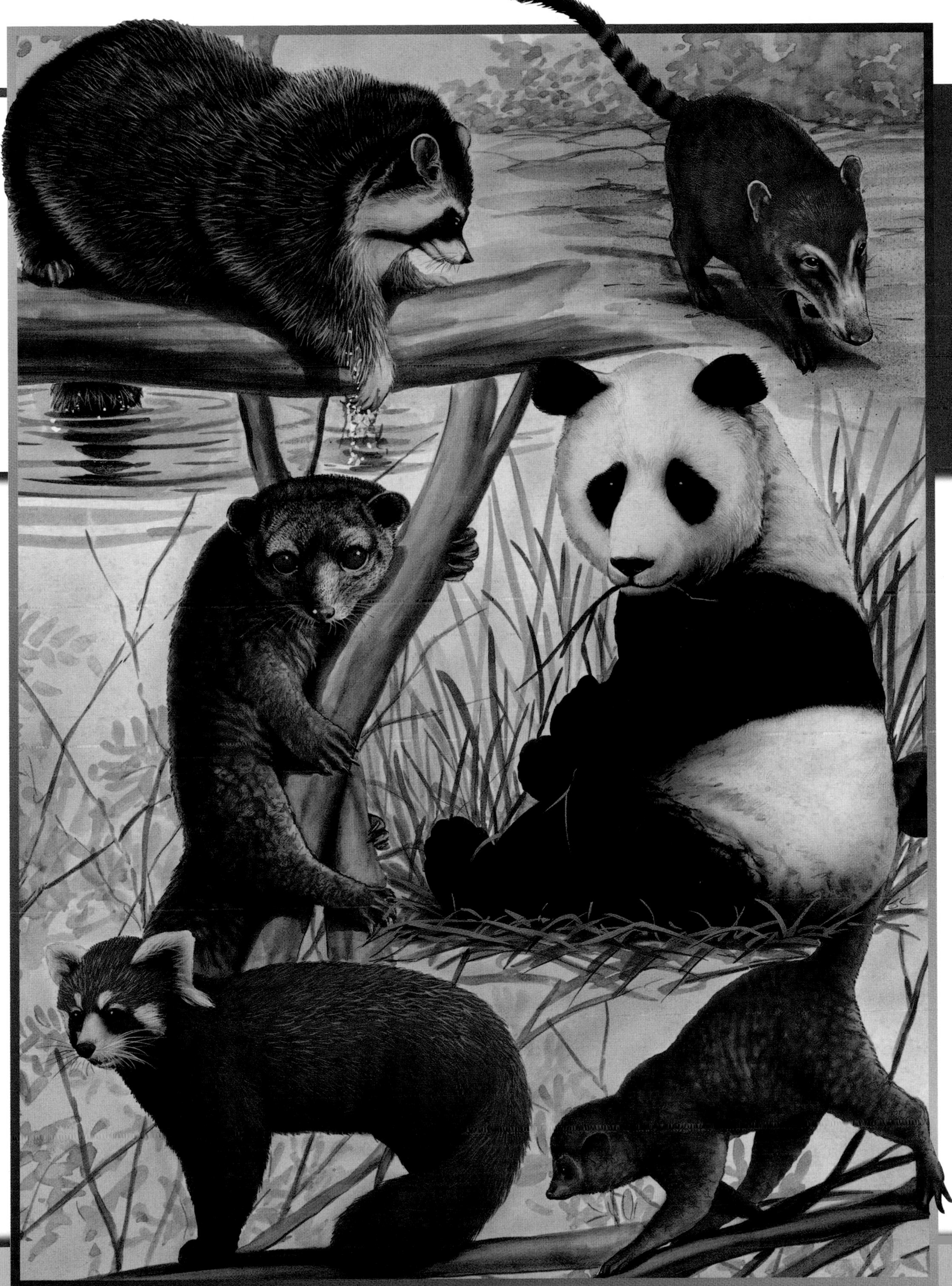

A European lake

- American mink
- Bleak
- Char
- Coot
- Emperor dragonfly (larva or nymph)
- Eurasian otter
- Great crested newt
- Great pond snail
- Kingfisher
- Marsh frog
- Marsh harrier
- Northern water vole
- Swan mussel
- Three-spined stickleback
- Vendace
- Water rail
- Water shrew

Freshwater life

Wildlife in the rivers and lakes of Europe differs from that in Africa and Asia.

The hidden world beneath the shiny ripples of a pond or river is just as fierce and deadly as a jungle. Some animals, including fish of course, stay submerged all their lives. But many others come and go between the land (terrestrial) and watery (aquatic) habitats. Otters, water voles, water shrews and coots swim under the surface to feed. The kingfisher dives in for a second or two to grab a scaly meal. Newts begin their lives in the water as tadpoles for a couple of months, and return there each spring to breed. Insect larvae (young) such as dragonfly nymphs may spend two or three years in the aquatic world, then emerge to become masters of the air.

Kingfisher

The common or Eurasian kingfisher is one of the smallest, fastest and brightest of the kingfishers. It plunges like an arrow into the water, stabs and catches a small fish in its dagger-like beak, and flies to a favourite perch. Here it bashes the flapping meal against a branch to make it still and easier to swallow.

Eurasian otter

This shy hunter can twist and turn underwater at amazing speed as it hunts its prey, mainly fish. It also eats freshwater crayfish and similar hard-shelled animals. It is still killed by people in some areas because it is thought to eat fish otherwise meant for anglers or human food. (See also page 52.)

Water rail

A tremendously shy bird, the water rail rarely comes into the open. It skulks among reeds, rushes and other bankside plants, searching for its food of worms, insects, other small animals and soft plants. However its loud call sometimes gives this bird away. It sounds like the squeaking squeal of a frightened piglet!

Great pond snail

The shell of the great pond snail may be up to 5 cm in length, making it one of Europe's largest snails. It usually eats bits of plants, animals and other decaying matter on the bottom of the pond. But it can also be a slow yet persistent predator and trap baby fish, tadpoles and even small newts.

Coot

Like the water rail and the moorhen (which has a red forehead rather than white), the coot has long, wide-splayed toes. These spread its weight to prevent sinking in soft mud or floating water plants. Coots dive in shallow water, tear bits off soft plants and carry them as they bob back up to the surface to eat.

Water shrew

The water shrew is huge among shrews, at 9 cm head-body length with a 5-cm tail. It swims in water to catch victims such as fish, frogs and worms. But it has to eat its own weight in food daily since it is such an active creature. So it often hunts along the bank too for grubs, slugs and spiders.

Birds of prey or ground hunters such as otters and mink rarely try to catch the kingfisher. Its bright feathers are warning colours which say: 'No! My flesh tastes really horrible!'

Northern water vole

Water voles are much larger than land voles, reaching 30 cm in length including the tail. This is soft and furry, and with the blunt head it distinguishes the water vole from the brown rat which is also a good swimmer. Water voles eat soft waterside plants.

Marsh harrier

As in many birds of prey, the female marsh harrier is larger than the male. Both are strong, powerful birds with wingspans of more than 100 cm. They glide low over riverbanks, lakesides and swampy regions on the lookout for frogs, voles, fish and small birds.

Char

The char (charr) is a close cousin of trout and salmon. It eats small animals such as freshwater shrimps and daphnias (water fleas). Char from different regions vary greatly, especially in size. In small lakes they reach 25 cm in length, in large rivers they may be twice as big.

Bleak

Only about 17 cm long, the bleak is a slim, darting fish from the carp group. It escapes its many predators by fast reactions and sudden speed. It swims in shoals near the surface, feeding on tiny animals and plants. It may rise to grab a fly.

Marsh frog

Europe's biggest frog, the marsh frog is also one of the loudest croakers. It can overpower mayflies, damselflies, slugs and many other victims. It returns to water in late spring to breed but soon leaves again for its life on land.

Swan mussel

The swan mussel is a bivalve – a mollusc type of shellfish similar to mussels, clams and oysters on the sea shore. It lives in lakes and slow, deep rivers. It draws a current of water into its shell, both for breathing and to filter out tiny bits of food.

Great crested newt

The long, fin-like crest on the back of this newt becomes taller and more colourful in spring when the male attracts a female for mating. His underside also glows bright orange with dark spots. This is a large newt, measuring 17 cm from nose to tail.

Three-spined stickleback

The three-spined stickleback only reaches about 9 cm in length. But it is an exceptionally tough and hardy fish. It is found in all kinds of water from farm ditches and stagnant ponds to the part-salty water of estuaries and coastal marshes.

A RIVERSIDE HOME

The riverbank or lakeshore is a valuable piece of 'real estate' for many creatures, from dragonflies to kingfishers and otters. Each individual or breeding pair competes for its own stretch which is known as a territory. The occupiers live and feed here, and keep out others of their own kind. They may leave droppings or scent marks, or make calls and noises, to warn others that the territory is occupied. For example, a water vole's territory is a ribbon of bank about 100–150 m long.

Emperor dragonfly

The nymph of this dragonfly is only 6 cm long but it is a giant terror in the underwater world of mini-beasts. It has a huge, hinged claw-like part called a mask on the underside of its head. This shoots out like a pair of pincers to grab tadpoles, young fish, pond snails and similar prey.

Swan mussels have very little in the way of interesting behaviour. They lie protected in their shells and move rarely and slowly. Yet they live to 20-plus years of age.

Dabbling, dipping and diving

- Barrow's goldeneye
- Cinnamon teal
- Common scoter
- Hooded merganser
- Ruddy duck
- Surf scoter
- Wigeon

Few rivers and lakes around the world are without ducks – the most common and familiar of all waterbirds (waterfowl). A duck spends much of its time swimming and so it has a light, buoyant body with a thick layer of waterproof feathers, and webbed feet to push itself along the surface or swim beneath. The dabbling or dipping ducks are the largest duck group. They dabble their beaks at the surface or 'up-end' so the head and neck are under the water with the tail pointing straight up into the air. Dabblers include the mallard, wigeon, gadwall and teal. Other ducks, such as the pochard and tufted duck, dive and swim deeper under the water to feed.

Hooded merganser
The male hooded merganser has a shiny green head and neck, and a fan-like crest. This American duck raises its young in tree holes.

Barrow's goldeneye
This diving duck breeds in North America and Iceland, in tree holes or rocky crevices. The male's glossy purple head makes his bright yellow eyes stand out.

Common scoter
Scoters are ducks of the open ocean, gathering there in winter flocks. They nest on small inland lakes but soon return to sea with their just-hatched chicks.

Ruddy duck
North and South America are the ruddy duck's original home. But it has been taken to brighten up the waterways of Britain and is spreading there. The ruddy duck has a stiff tail held up at an angle. The male's courtship display involves rapid paddling as he holds his head against his chest.

Wigeon
Wigeons are often seen flapping and wheeling in flocks over wetlands, making their characteristic whistling calls. These ducks feed more like geese, pecking at plants as they waddle through meadows and marshes. The male has delicately patterned plumage with a pale cap or crown on the head.

Cinnamon teal
This duck lives on marshes, lakes and ponds in western North America. It is named from the glowing, rusty-red shades of the feathers on the head and body of the male, or drake. He also has bright, striped-looking wings folded along his back. But the female, as in most ducks, is dull mottled brown.

Surf scoter
The surf scoter is larger and has a thicker, stronger beak compared to the common scoter. It is a North American duck and has distinctive white patches on its head. Surf scoters nest in the northern woods of Canada. Then, as their name suggests, they move to coastal seas for the rest of the year.

Many ducks quack but the hooded merganser is one of the world's quietest birds. Only the male makes a noise, which is a throaty purr like a cat – and then only at breeding time.

The biggest family of fish

- Bigmouth buffalo
- Dace (dart)
- Goldfish
- Minnow
- Nase (sneep)
- Shorthead redhorse
- Stone loach
- Pike (northern pike)

The carp family is the largest of all fish groups with some 2000 different species. They live in fresh water and are mostly strong, deep-bodied and eat small food items such as bits of plants and water-living grubs. Different kinds of carp live mainly in the northern parts of the world. They have been taken to many other regions as food fish and for anglers. The common carp is now found almost worldwide and is an especially powerful and wily fish. In addition goldfish, koi, mirror carp, golden carp, leather carp and many other varieties have been bred as ornamental fish for ponds and lakes.

Dace
A quick and darting fish, the dace rarely grows longer than 25 cm. It likes clean fairly fast rivers and eats flies, grubs and other small animals.

Minnow
Many small or young fish are 'minnows' but the minnow is also a distinct kind or species. It is only 10 cm long and a common victim of bigger fish.

Nase
A fish of fast and gravel-bottomed rivers, the nase lives in Europe and Western Asia. It scrapes small plants off the stones with its hard, horny lips.

Pike
The pike is not a member of the carp group – but it does eat carp. It is a powerful predator up to 100 cm long with a mouthful of sharp teeth. It dashes out of water plants to ambush its prey.

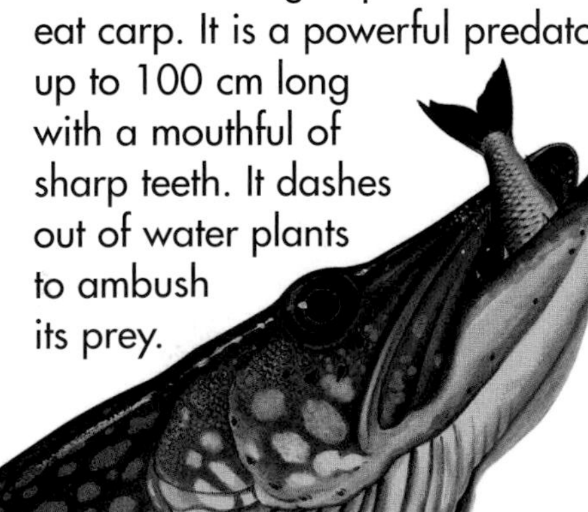

Bigmouth buffalo
This massive, deep-bodied fish reaches 100 cm in length and lives in large lakes and rivers in eastern North America. It feeds on water plants and animals such as pond snails.

Goldfish
Wild goldfish live in weedy ponds and lakes across Central and Eastern Asia. They grow to 30 cm and have been bred in hundreds of colours, sizes and varieties for the aquarium.

Shorthead redhorse
Redhorses are types of sucker carp, named from their big, fleshy lips. They feed mainly on water insects, worms and grubs.

Stone loach
This small loach, only 15 cm long, lies camouflaged on the river or lake bed by day. It grubs among the stones for small worms, shellfish and similar food.

During the spring breeding season a female bigmouth buffalo lays up to half a million eggs.

Spear-fishing for a meal

- Black-capped kingfisher
- Cuckoo roller
- Green wood-hoopoe
- Jamaican tody
- Malachite kingfisher
- Pied kingfisher
- Scaly ground roller
- Stork-billed kingfisher
- Turquoise-browed motmot

With their bright plumage and rapid darting flight, kingfishers are well known but rarely seen. If disturbed they flash away with a whir of wings. Most feed on fish, as their name suggests, but some take insects, lizards, frogs and similar prey. The bird plunges like an arrow into the water and grabs the victim in its spear-like beak. Kingfishers nest in holes. Some dig their own tunnels in soft riverbanks. Others use hollow trees. Wood-hoopoes live only in Africa and use their long, curved beaks to gather insects from tree bark. Like the motmots of Central and South America and the todies of the Caribbean, they are close cousins of kingfishers.

Stork-billed kingfisher
This is one of the largest and most powerful kingfishers, reaching a bill-tip to tail-end length of 36 cm. It lives along wooded streams, lakes and paddies (flooded rice fields) in India and Southeast Asia. As well as catching fish, it includes insects, lizards and small birds in its diet.

Jamaican tody
Todies are tiny birds with mainly bright green plumage. Their hunting method is to sit and wait on a branch, then quickly fly into the open to grab a butterfly, beetle or similar insect flying past. Todies can hover in mid air almost as well as hummingbirds. This allows them to pick insects off leaves.

Black-capped kingfisher
A small kingfisher of river banks, rice fields and mangrove swamps, this species is found from India across Southeast Asia to the Philippines.

Malachite kingfisher
Common in parts of East and Central Africa, the malachite kingfisher's dazzling blue upper parts are set off by its long red beak.

Turquoise-browed motmot
This motmot lives in the steamy forests of Central America. Spoon-shaped tail feathers are a feature of the motmot group and are used to signal to other birds.

Cuckoo roller
This bulky, strong kingfisher is found in the forests and scrub of Madagascar. It feeds mainly on large insects and lizards high in the treetops.

Green wood-hoopoe
Wood-hoopoes are busy birds that cackle constantly as they flap through the trees. They poke their long bills into bark for insects.

Pied kingfisher
The pied kingfisher of Africa and Asia hovers over open water before plunging after fish. Pied kingfishers breed in groups and, unusually for birds, help each other to raise their chicks.

The biggest kingfisher rarely eats fish – but it does 'laugh' very loudly. It is the kookaburra of Australia and it preys on all kinds of small animals, even rats.

Silvery spotted leapers

- Atlantic salmon
- Dipper (a river bird)
- Leech
- Northern grayling
- River lamprey (brook lamprey)
- Trout (sea trout)
- Steelhead trout

Leaping waterfalls and rapids, on the way up their home river to breed, salmon are among the world's best-known fish. (And the tastiest.) Along with trout, pike, charr and smelt, they make up the large salmon family with more than 500 different species. Most live in northern parts of the world and are predators, hunting smaller animals for food. Many, like salmon and sea trout, are also migratory. They grow up in a river for a few years, journey out to sea for several more years, then return to the same river to spawn (lay eggs). They probably find their way by 'smelling' the exact mixture of chemical substances in their home stream.

Atlantic salmon

Salmon spend from 2 to 6 years in their home river, then head out to sea where they grow up to 1.5 m long, powerful and fast as they feed on smaller fish. After between 1 and 4 years at sea they head back upriver to breed in the gravelly stream where they hatched. Most then die but some make the journey twice.

Trout

Few fish are as widespread as the trout, which has been taken to all continents for angling and as food. The variety called the brown trout stays in a lake or river all its life. The sea trout (shown opposite) is more silvery and has a life cycle like the salmon, heading out to sea and then returning to breed.

River lamprey

Lampreys are not members of the salmon family but very strange fish with an almost prehistoric body design. They lack jaws. The mouth is a round sucker edged with tiny teeth. The lamprey usually lives as a parasite. It sticks onto a larger fish, its host, and rasps its way through the skin to suck its blood and body fluids.

Northern grayling

The grayling looks like a small trout, about 45 cm long, but with a larger sail-shaped back or dorsal fin. Like most members of the salmon family it has little, sharp teeth. Grayling live in Northern Europe and Northwest Asia and feed on small water creatures such as worms and insects.

Steelhead trout

Steelheads show the typical feature of the salmon family – the small lobe-like adipose fin on the upper rear body, between the main dorsal fin and the tail. The fish's blue-grey head looks like polished metal. The steelhead trout shown opposite is about to tackle a large leech on the stony river bed.

A salmon loses up to half its body weight as it battles its way against the current to the stream where it grew up.

Royal birds of open waters

- Australian black swan
- Black-necked swan
- Mute swan
- Trumpeter swan
- Whistling swan
- Whooper swan

Swans are the largest waterfowl and some of the heaviest flying birds. They need a long stretch of open water to take off, running across the surface as they pick up speed. Once in the air they are powerful fliers and some types cover huge distances each year on migration. A swan has a long, flexible neck to stretch almost 100 cm down into the water when feeding. Because of the swan's large size, strength, beauty and grace it has featured in many myths and legends. Swans have also inspired artists, writers and musicians for centuries and are often called 'royal' birds. Some types, like mute swans, are tame and live on busy lakes and rivers.

Whooper swan Whooper swans are very similar to trumpeter swans except for the bright yellow triangular patch on the beak. They live across Europe and Asia, breeding in northern areas like Iceland, Scandinavia and northern Russia. They migrate to spend the winter farther south in Western and Eastern Europe.

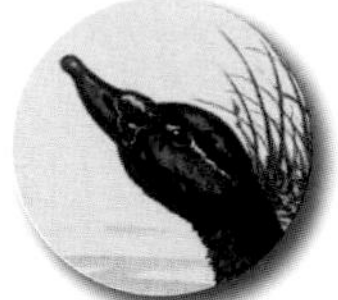

Black-necked swan This swan is found in South America and is particularly common in Argentina and Brazil. It prefers marshes and shallow water to deeper rivers and lakes. Flocks of black-necked swans sometimes rest on the sea just offshore, bobbing up and down one behind another in a long line.

Whistling swan Whistling swans are the North American variety of the swan species called the tundra swan. Another variety of the same species is Bewick's swan of Asia and Europe. Both varieties are small for swans. Yet they fly huge distances from their Arctic breeding places for winter on coasts and marshes to the south.

Australian black swan Most swans are white. Australian black swans are not. They have been taken to lakes and ponds around the world as a dark, contrasting addition to the local white swans. In their original home of Australia black swans breed in large colonies and often form huge flocks.

Mute swan Mute means silent, but the mute swan is not. It makes various grunts and hissing noises. However it does not have the musical calls of many other swans which are named after the sounds they make, like the whistler and trumpeter. In flight the wings of the mute swan beat powerfully and slowly with a swishing sound.

Trumpeter swan A North American swan that certainly lives up to its name, its loud, bugle-like cries carry over great distances. Trumpeters breed mainly in the far north of North America, especially Alaska. They move south for the winter but not very far, to the open water along Canadian coasts and on large lakes in national parks.

The mute swan is one of the heaviest of all flying birds, weighing up to 18 kg. It usually needs at least 50 m of open water for take-off and landing.

Strange survivors from prehistory

- Bowfin (grindle or Great Lakes dog-fish)
- Longnose gar (long-nosed gar-pike or garfish)
- Spotted gar (spotted gar-pike or garfish)

The vast majority of fish belong to one enormous group, the bony fish. They have skeletons made of bone, not cartilage like sharks and rays. But among the 20,000-plus kinds of bony fish are several small groups which resemble their prehistoric cousins, almost like 'living fossils'. One is the bowfin group. It's so small that it has only one member – the bowfin. This fascinating fish lives in pools and streams in north-east and central North America. Fossils of its ancient cousins dating back to dinosaur times have been found across Europe and Asia. Gars were also once widespread. Now only seven kinds (species) survive in North and Central America.

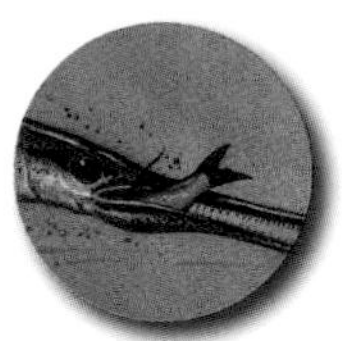

Longnose gar

The longnose gar (shown opposite eating a threadfin shad) is a lurking predator like the other gars. It grows to about 1.7 m in length and lives in lakes and rivers throughout North America. Gars tend to wait among water plants, or alongside roots or branches, and dash out with a rush to grab their prey.

Spotted gar

The spotted gar has ideal camouflage for hiding among plants or bits of sunken wood. A gar's long mouth has many sharp teeth and the slender jaws can be flicked sideways at speed through the water to snap up victims. The anal (underside) fins and dorsal (back) fins are placed to the rear, near the tail, for bursts of speed.

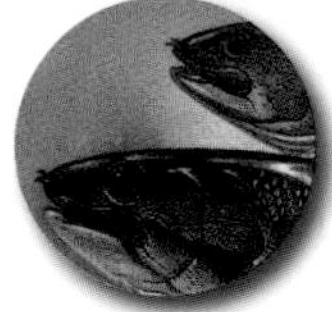

Bowfin

This unique fish has many unusual features. It is named after its long back or dorsal fin which has a curve like a longbow. It also has a rounded rather than forked tail, and an almost rod-like shape with a blunt head and deep, wide body. It grows to about 100 cm in length. Bowfins are fearsome hunters of smaller fish, crayfish, freshwater shrimps, frogs and similar prey.

The male bowfin is usually slightly smaller than the female and he has a dark spot edged with yellow or orange at the base of his tail. He makes a shallow bowl-like nest on the bottom by biting away plants and swishing away mud and stones with his tail. After the female lays her eggs there he guards them fiercely. He also continues to defend the babies when they hatch.

FISH OUT OF WATER

Bowfins and gars have swim bladders which can work like lungs to breathe air, as in the lungfish (see page 120). This allows the fish to survive out of water for a day or more. Breathing air is a useful feature for fish that dwell in warm, still, stagnant water, often found in tropical marshes and swamps. This water has very little oxygen dissolved in it. So the fish obtains extra supplies by gulping air.

Another strange feature of the gars is their scales. These are diamond-shaped, thick and slab-like. They are called ganoid scales and give good protection, like armour. But they are much heavier than normal fish scales. They also limit movement because the body can bend or flex less when swimming.

The male bowfin is the most dedicated father of all fish. He protects his young until they grow to about 10 cm long.

Desert by night and day

- Antelope jackrabbit
- Banded gecko
- Bighorn
- Broad-billed hummingbird
- Cactus wren
- Chuckwalla
- Collared peccary
- Coyote
- Desert kangaroo rat
- Diamondback rattlesnake
- Elf owl
- Gambel's quail
- Gila monster
- Gila woodpecker
- Harris hawk
- Hog-nosed skunk
- Lesser long-nosed bat
- Magnificent hummingbird
- Mule deer (black-tailed deer)
- Ringtail (cacomistle)
- Roadrunner
- Side-blotched lizard
- Swift fox (kit fox)
- Tarantula
- Texas horned lizard

Coping with drought

Desert and very arid (dry) scrub cover the south-west corner of North America.

The desert by day is a hot, dry and fairly empty place. A few hardy animals brave the blazing sun to munch at cacti or thorny scrub, or snatch small prey. They rest during the middle of the day in any shade they can find, to avoid the worst of the heat. Even 'cold-blooded' animals such as lizards, snakes and insects must be careful. Their body temperatures follow the temperature of the surroundings and they can overheat. But as the sun sets and the air cools and the dew falls, the desert comes alive. A whole new batch of animals is out and about. They emerge from their nests, burrows and tunnels to look for food and perhaps mates at breeding time.

Elf owl

Most owls roost by day in holes inside tree trunks. There are few trees in the desert so the tiny elf does the same in a hole inside a giant saguaro cactus or similar desert plant. It usually takes over an unoccupied hole made previously by a woodpecker. As night approaches the elf owl swoops across the desert to snatch prey from the ground with its claws. It feeds mainly on insects such as grasshoppers, crickets, beetles and moths. It may also take small snakes and lizards, spiders and even scorpions, although it pecks or tears off the stinging tail first. Elf owls live in the dry regions of south-west North America. The male is a busy father, feeding the mother as she sits on the eggs and also her and the chicks when they hatch.

Diamondback rattlesnake

This rattler has an extremely poisonous bite. The 'rattle' is made of loose, collar-shaped scales joined like links in a chain. A new one is added each time the rattler sheds its skin in the usual snake way, once or twice yearly. But old links fall off so rattle size is not a reliable guide to age.

Hog-nosed skunk

Like many desert animals, the hog-nosed skunk is an opportunistic feeder. This means it takes a wide range of food, from fruits and berries to grubs, worms, lizards and snakes. Being an adaptable eater is a great aid to survival in the desert. This skunk is about 70 cm long including the tail.

Cactus wren

Most wrens are small birds and feed on tiny insects. The cactus wren is much larger, 20 cm from beak to tail, and feeds on grasshoppers, wasps, big beetles, and even frogs and mice. It prefers running or hiding to flying. It builds its large, domed nest among the thorns of a cactus or similar desert plant.

Antelope jackrabbit

The huge ears of this desert hare can both hear the tiny sounds of an approaching predator, and give off excess body warmth to keep the creature cool. Its long deer-like legs give enormous speed. During the drought it braves the thorns of cacti and yuccas to nibble their fleshy parts for moisture.

The elf owl is one of the smallest owls in the world. It measures just 13–14 cm from beak to tail – smaller than an average human hand.

Mule deer

There are several varieties of this deer in western North America. They stand about 1.2 m tall at the shoulder and the bucks (males) have impressive antlers by four years of age. They do not form herds like most other deer but live in small family groups.

Harris hawk

Also called Harris's hawk, this powerful and fast-flying bird of prey soars over the desert by day. Its large eyes can spot prey such as jackrabbits, snakes and lizards from 2–3 km away. As in most hawks and buzzards, the female is larger than the male.

Coyote

The long, mournful night-time howl of the coyote is a famous sound of a lonely creature in the wilderness. But recent studies show that coyotes sometimes form packs like other dogs. They eat a vast range of prey from beetles to deer as well as fruits. (See also page 196.)

Ringtail

This creature looks like a combination of raccoon, fox, cat and stoat or mink. It also has a huge range of local names including raccoon-fox, banded-tailed cat, cat-squirrel, and further south in Mexico, cacomixtle ('rush cat') and tepemixtle ('bush cat'). In fact it is a close cousin of the raccoon. This lightweight, agile creature is about 80 cm long including the bushy tail and is an expert climber. Ringtails sleep by day in dens among the rocks, branches or plant roots. They eat all kinds of small animals like mice, rats, lizards, birds and insects, and also plenty of fruits.

Lesser long-nosed bat

This bat is not an insect-eater like most of its relatives, but a nectar-feeder. It roosts (sleeps) by day in caves, and flies out at night in search of open flowers. It visits plants such as organ-pipe and barrel cacti.

Chuckwalla

The chuckwalla is a strong and tough-looking lizard about 40 cm long. It sleeps at night in a cave or rocky crevice and comes out to bask in the morning sun, warming up for the day's activity. It feeds on all kinds of plant food such as flowers, leaves, buds and shoots.

Texas horned lizard

This lizard looks fierce but is only little, about 15 cm in total length. It eats little prey too, mainly ants. The pointed scales along its sides and around its head give good protection against predators. Despite her small size, the female lays up to 35 eggs.

PLANNING AHEAD

Many desert animals store food and moisture in their bodies, building up reserves during the short time of plenty. The chuckwalla lizard can store water in special glands under the folds of skin along its neck and sides. As it uses up this water its body becomes thinner and more wrinkled. Mice and rats hoard seeds in their burrows. The gila monster lizard stores excess food as fat in its plump tail. It can use this both for nourishment and body water – just like the camel's hump.

Tarantula

Three main groups of spiders are known as tarantulas. One is the 'original' tarantula, small wolf-spiders of Italy. Another is the large bird-eating spiders of South America. Third is the large hairy spiders of the deserts in south-west North America. They hunt smaller creatures – including little spiders.

The chuckwalla is one of the few plant-eating animals that can feed on the creosote bush. The natural toxic (poisonous) chemicals in this bush kill most herbivores.

Our constant cousins

- African pygmy mouse
- Black rat
- Brown rat (Norway or water rat)
- Spiny mouse
- Desert jerboa
- Harvest mouse
- House mouse
- Rock mouse
- Striped field mouse
- Three-toed jerboa
- Wood mouse
- Yellow-necked field mouse

It is said that, wherever you are in the world, you are never more than a few metres away from a mouse or rat – perhaps even at sea! There are well over 1200 different kinds, or species, of these rodents so they make up by far the largest group of mammals. Many of them are small, quick and adaptable and have taken to living near people. They eat our stores of grains and food in our cupboards and larders, just as they feed on seeds, nuts and berries in the wild. Rats and mice have long, sharp front teeth to chew and gnaw even the hardest items. Some mice are as small as your thumb. The largest rats are almost the size of pet cats.

House mouse
House mice have lived in buildings and 'shared' our food for thousands of years. They breed quickly too. A female can produce 25 young in six months.

African pygmy mouse
The pygmy mouse lives in grassland and scrub in East Africa. It is one of the smallest mice, measuring just 10 cm in total length.

Striped field mouse
Unlike most mice, this striped species eats few green plants. It feeds on seeds and fruits and also catches insects, worms and slugs.

Harvest mouse
An expert climber, this tiny mouse uses its long, grasping tail to clamber among grass stems. It lives in a tennis-ball-sized nest of woven stalks.

Desert jerboa
Jerboas are like tiny kangaroos. They make huge leaps using their long back legs, balanced by the long tail. They live in African and Asian deserts.

Rock mouse
This large mouse lives in dry, rocky places and makes its nest in a safe crevice. Its long whiskers help it to feel its way among the stones at night.

Yellow-necked field mouse
This agile mouse is common across Europe. It lives mainly in woods, fields and gardens. Like many mice it is active at night with big, beady eyes to see in the dark.

Brown rat
The brown rat lives almost everywhere, including farms, gardens, rubbish dumps, city centres, ditches and even sewers. It is an excellent swimmer and sometimes called the 'water rat'.

Black rat
Black rats live mainly in warmer regions. They are also called ship rats because they can climb mooring ropes into boats and also swim well. They carry diseases such as plague.

Some jerboas can jump 3 m – equivalent to a person leaping six times the human long-jump world record.

Silent flight, hunters at night

- African wood-owl
- Barking owl
- Burrowing owl
- Elf owl
- Malay eagle owl
- Papuan hawk-owl
- Seychelles scops owl
- White-faced scops owl

Owls are the specialist night hunters of the bird world. They catch mainly small animals like mice, voles and lizards, although larger owls prey on rabbits and birds (including other owls) while smaller types hunt moths, beetles and other little creatures. The fishing owls of Africa and Asia grab fish, frogs and crayfish from shallow water using their long, unfeathered legs and sharp 'fish-hook' claws. The owl's wing feathers have soft edges which means they are almost totally silent in flight, so the owl can swoop undetected on its prey. Huge eyes give the owl good night vision, but its hearing is even better – four times more sensitive than the ears of a cat.

African wood-owl
This medium-sized owl is common in south and east Africa. But unlike many owls that hunt at dusk, it only comes out in darkness, and so it is heard but rarely seen.

Barking owl
The growls and barks of this owl are eerily dog-like. In New Guinea and Australian forests it eats opossums as well as the usual owl prey.

Papuan hawk-owl
This rare owl lives in New Guinea forests where logging is a major threat. It has a hawk-like long tail and rounded wings.

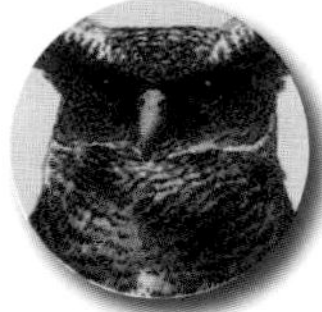

Malay eagle owl
Eagle owls are among the largest owls. Some have wingspans of 1.5 m. The Malay eagle owl is slightly smaller but still a powerful hunter, recognized by its very long 'ear tufts'. (These are not real ears but simply long feathers.) Its mysterious hooting, groaning calls are said to be made by demons of the night!

White-faced scops owl
The ghostly black-rimmed white face and long 'ear tufts' of this African owl make it one of the most distinctive members of the family. Its favourite hunting method is to sit and wait on a tree branch, then drop silently onto a passing victim below – a mouse, an insect or even a scorpion.

Elf owl
Only 12 cm long, this tiny owl from southern North America is one of the smallest owls. In its desert habitat it nests in old woodpecker holes in the giant saguaro cactus.

Seychelles scops owl
Known only from one island in the Seychelles, this small owl inhabits old forests on mountain slopes. It has a call like a clock's tick-tock.

Burrowing owl
Small and long-legged, this owl lives in grasslands throughout the Americas. It digs its own burrow or takes over one from a rabbit or prairie dog.

The barn owl is the most widespread bird in the world. It is found on every continent (except Antarctica) and in almost every habitat from remote mountains to busy towns.

Stealthy, silent, smaller killers

- Black-footed cat
- Caracal (African lynx)
- Lynx (northern lynx, European lynx, American lynx)
- Leopard cat
- Pallas's cat (manul)
- Puma (cougar, mountain lion, also called panther in parts of its range such as Florida)
- Serval

The 'small' cats are not all small. Largest is the puma which can grow to a head-body length of 1.8 m and weigh 100 kg. This is much larger than the smaller types of big cats such as the snow leopard (see pages 94 and 146). However most of the 28 kinds or species of small cats are in the medium-to-little group, slightly larger than a big pet cat. They live solitary lives, stalking alone and at night or in twilight for suitable-sized victims. Most live in forest, scrub and bush. But the caracal can cope with the searing heat of the Sahara desert while the thick-furred lynx ventures onto the snowy, treeless tundra of the far north after its chief prey, snowshoe hare.

Caracal

The caracal can cope with a wide variety of conditions from desert to grassy plains and bush. This cat hunts all kinds of animals from mice and lizards to deer, using speed to catch prey like hares. Sometimes it attacks farm animals such as goats and chickens. Caracals live in Africa and Southern Asia.

Puma

Few cats are as adaptable as the puma. It dwells in many types of habitats, from snowy mountains to subtropical swamps, and ranges from southern Canada through North and Central America to Patagonia in South America. It has a muscular build and uses mainly the stalk-and-pounce method.

Serval

The serval of Africa is a smaller version of the cheetah with a spotted coat, slim body and long legs built for speed. Unusual for a cat, it is often on the prowl by daylight, and it sometimes eats fruits and flowers as well as its usual food of prey animals. It may leap 3 m in the air to knock down a low-flying bird with its paw.

Leopard cat

This small cat likes thick, swampy forest or scrubland. It is a good swimmer and an excellent climber, waiting in the branches to drop silently onto prey passing below. Leopard cats are named after their spotty coats and inhabit many islands of Southeast Asia, north to the Philippines and Japan.

Pallas's cat

Mountains, rocky areas, scrubby foothills and grassy uplands across Central Asia are home to this sturdy cat. It has short legs, small ears, and longer fur than any other type of small cat. It preys mainly on rats, mice, marmots, rabbits, pikas and similar small mammals, and also on ground-dwelling birds.

Black-footed cat

This small cat is so little that it has trouble tackling a big rat. It catches mainly mice, voles, small snakes and lizards, and insects such as locusts and beetles. It is found mainly in grassy, rocky areas in Southwest Africa, and is named from the black patches on the undersides of its lower legs and feet.

The smallest small cat is the black-footed cat. Its head and body are 35 cm long, its tail about 15 cm, and it weighs 1–2 kg. Many pet cats are larger.

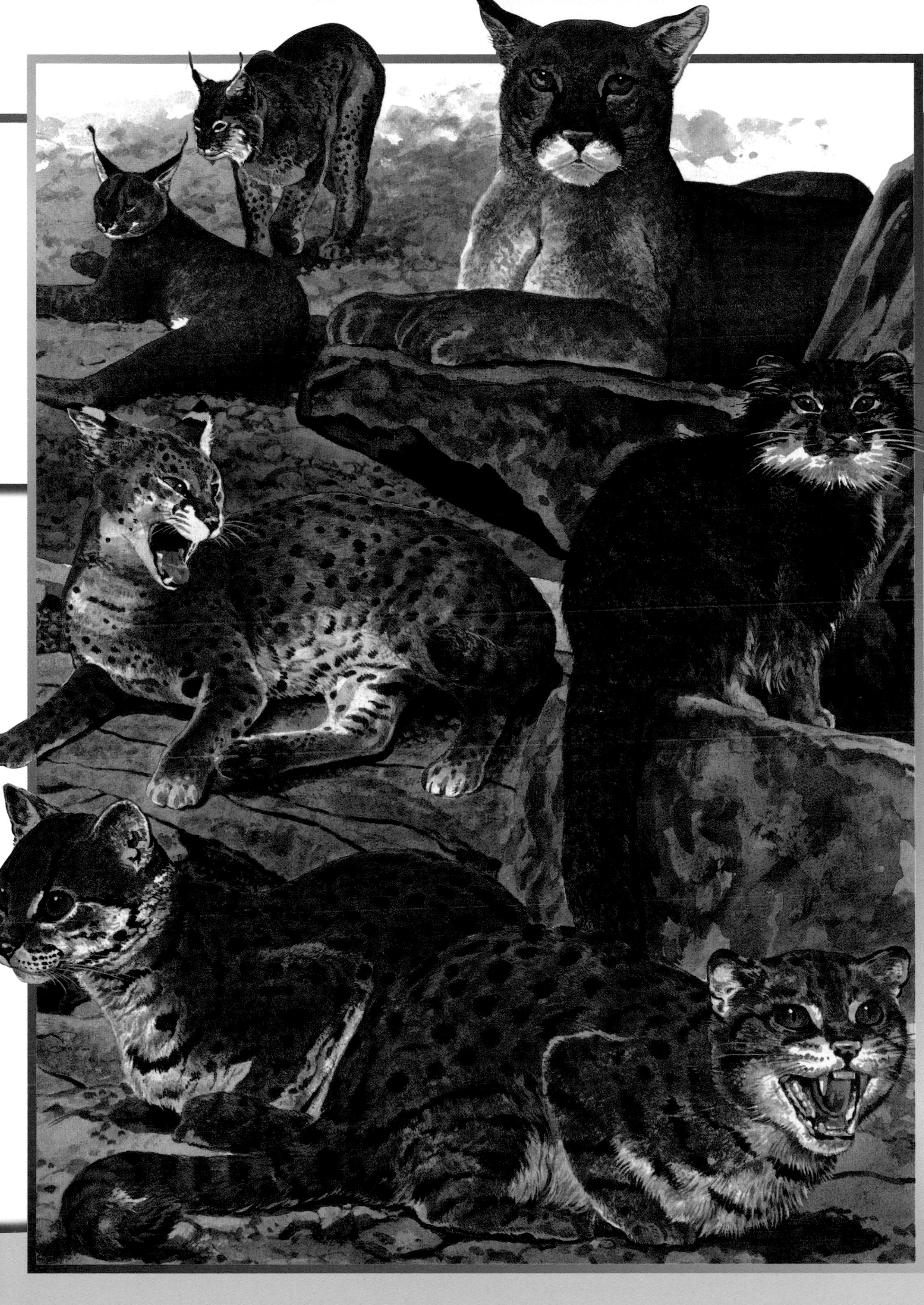

Kings of the bird world

- Australian little eagle
- Bonelli's eagle
- Crested serpent eagle
- Philippine (monkey-eating) eagle
- Spanish imperial eagle
- Steller's sea eagle
- White-bellied sea eagle

Eagles are among the biggest, strongest and fiercest birds in the world. Like other raptors (birds of prey) they have powerful feet and talons (claws) for seizing victims and sharp, hooked beaks for tearing flesh. Their large eyes can spot prey many kilometres away. But most eagles are difficult to spot themselves because they are wary and live in remote places like islands, mountains and thick forest. Smaller eagles take mice and voles. Medium-sized types swoop on rabbits. Fish and sea eagles have extra-curved talons and rough toe skin to grasp their slippery meals. The largest eagles carry away monkeys, small deer and hares.

Steller's sea eagle
One of the most powerful of all birds, this eagle has a wingspan of 2.4 m and a massive beak to rip the flesh from fish, dead seals and beached whales. Steller's sea eagles breed around the Bering Sea between Asia and North America. They often gather off the island of Hokkaido, Japan.

Crested serpent eagle
This small eagle is widespread across Southeast Asia. But it is seldom seen since it tends to hunt from the cover of the forest, suddenly swooping from a tree to surprise its victim. As the name suggests, this eagle has a head crest of feathers and it hunts snakes and lizards as well as the usual prey.

White-bellied sea eagle
This small sea eagle lives from India across to Australia. It soars over the water and glides down to grab a fish just under the surface.

Australian little eagle
The little eagle of Australia and New Guinea has long, feathered legs and a short crest on its upper neck. It hunts a wide range of prey including worms.

Spanish imperial eagle
A rare bird from central and southern Spain, this eagle favours mixed habitats with forest, scrub and marsh or lakeside.

Bonelli's eagle
Forest and scrub for hunting, with cliffs nearby for nesting, are the favoured home of Bonelli's eagle. It soars over hillsides and valleys, watching carefully for prey below – mainly small mammals and ground birds. Sadly Bonelli's eagle is becoming rare. It still survives in remote parts of Spain.

Philippine eagle
Truly huge at 90 cm from beak to tail, this great bird equals the harpy as the world's biggest eagle. Despite its size and power it is agile enough to pluck monkeys from tropical forest tree tops. A threatened species, only a few hundred of these eagles survive in the mountains of certain Philippine islands.

The bald eagle, the national bird of the USA, is a kind of fish eagle – and is not bald at all. However from a distance its white head feathers make its head look bare.

Secret squeakers of darkness

- African yellow-winged bat
- Hammer-headed bat (horse-headed bat)
- Hoary bat
- Lesser horseshoe bat
- Long-eared bat
- Mexican fishing bat
- Noctule bat
- Red bat
- Spotted bat

Few people see bats. They fly at night, mostly in dense forests. Yet of the 4080 or so species of mammals, nearly one in four is a bat. Bats are found in all but the coldest parts of the world. Most are hand-sized and feed on small flying creatures such as moths. However some catch birds, mice or even fish. The large day-active bats known as flying foxes feed on fruit. The bat's arms are wings consisting of thin skin stretched between the enormously long hand and finger bones. A bat finds its way in darkness by making high-pitched clicks and squeaks and listening to the echoes which bounce back off nearby objects. This is called sonar or echolocation.

Noctule bat
The noctule is one of the largest European bats. It flies high, fast and straight as it hunts for insects. It also migrates up to 1500 km between its summer and winter homes.

Hammer-headed bat
A bat of tropical Africa, only males have the donkey-like nose. More than one hundred males may gather and fly in a swarm to attract females at breeding time.

African yellow-winged bat
This large bat has huge ears and yellow or orange wings. It hunts from a perch by day and night, and eats reptiles, birds and fish in addition to insects and spiders.

Mexican fishing bat
This bat detects the ripples made by a rising fish. It then swoops low over the water surface and hooks the fish in its long, sharp claws.

Long-eared bat
The big ears help to collect squeaky echoes as this bat navigates through woods in the dark. It is a skilled hunter of night-flying insects such as moths.

Hoary bat
The hoary bat is a fast and powerful flier. It has spread from mainland America to the Hawaiian Islands in the Pacific, helped by following winds.

Red bat
Most mother bats have one or at most two babies. The red bat often has three. Like many other bat species, red bats hibernate inside hollow trees.

Lesser horseshoe bat
Horseshoe bats are named after the U-shaped folds of skin around the nose, which help to direct their high-pitched (ultrasonic) squeaks.

Spotted bat
This bat is found in Mexico and the southern states of the USA. Like many bat species it is a communal rooster – it gathers in groups or colonies to rest.

The Mexican free-tailed bat forms the largest colonies of any bat (and of almost any animal). As many as 10 million may cluster together in a single cave.

At the waterhole

- African buffalo
- African elephant
- Avocet
- Blue wildebeest (gnu)
- Carmine bee-eater
- Cattle egret
- Common (plains) zebra
- Egyptian goose
- Giraffe
- Great white egret (common egret)
- Griffon vulture
- Haartebeest
- Hammerkop
- Hippopotamus
- Lion
- Nyala (female and male)
- Thomson's gazelle
- Warthog

On the savannah

Vast savannahs and open bushland cover much of East and Southern Africa.

The wide-open grasslands of Africa are home to some of the world's most spectacular wildlife. The largest of all land animals live here – elephants, rhinos, giraffes and hippos. Grasses grow because the climate is slightly too dry for forests or woods, but slightly too moist for scrubby semi-desert. A few scattered trees and bushes give valuable shade during the long, hot, dusty dry season. As the waterholes and rivers shrink, life becomes more perilous. The many risks include drought, starvation, bushfire and the ever-present predators and scavengers. Then there's a distant clap of thunder, the skies darken, the rains pour down and the plains turn green again.

Griffon vulture
These vultures roost, nest and search for food in groups, which is unusual for this type of bird. They live entirely by scavenging. (See also page 100.)

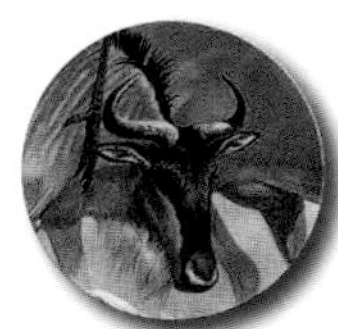

Blue wildebeest
In the dry season huge herds of wildebeest (gnus) trek more than 1500 km to find fresh grazing and holes which still hold water. (See also page 106.)

Haartebeest
As the land dries out and food supplies dwindle, these large, strong antelopes form mixed herds with zebras and other grazers.

Carmine bee-eater
This large, bright bee-eater always feeds and nests in groups. It lives mainly in North and Central Africa. (See also page 162.)

Nyala (male)
This is a large male antelope, with a head and body up to 2 m long. Nyalas usually stay in thickets or among bushes and emerge only to find water.

Nyala (female)
Smaller than the male and lacking horns, the female nyala is also much lighter reddish-brown. These antelopes eat tender young grass and tree leaves.

Lion
The male lion has a head and body about 2 m long. Its tail is 100 cm in length and like many savannah animals it has a tufted end which acts as a useful fly-whisk. (See also page 94.)

Warthog
A tough and fierce wild pig, the warthog has large tusks which are its extra-long canine teeth. It eats all kinds of plant foods and even scavenges on carrion. (See also page 164.)

Common (plains) zebra
A zebra stallion (male) takes over a herd of females at 7–8 years of age. He may last up to 10 years, fathering all the foals, before a younger male replaces him.

A full-grown African buffalo has been seen to fight off a group of five lionesses.

African buffalo

These buffaloes are big, powerful and aggressive animals. They do not hesitate to charge and even a lion pride is wary of them. They rest by day in long grass or wallow in the mud at the water's edge. At night they feed on grass and the young leaves of trees and bushes.

African elephant

Elephants prod and gouge the ground with their tusks to dig up roots and tubers, and also to encourage water to seep from the soil. In this way they actually create new waterholes. A baby elephant is fed by its mother for 2 years. (See also page 96.)

Giraffe

The world's tallest animal, the giraffe can reach leaves more than 6 m from the ground. Giraffes live in herds or troops of about 6–10 which consist of a chief male plus females and their calves (young). In the dry season these may gather to form larger groups. (See also page 98.)

Thomson's gazelle

This is one of the smallest, daintiest and fastest gazelles. It grows to only 100 cm head-body length and is a main food item of many savannah predators such as lions, leopards, hyaenas, jackals and especially cheetahs. It lives in loose herds of up to 500.

Hammerkop

This strange bird with its rear head crest is a close relative of storks and flamingoes. It stays near rivers and pools to catch fish, frogs, crayfish and grubs. The hammerkop measures 50 cm from beak to tail. It lives across Africa and in the Middle East. (See also page 118.)

Cattle egret

This type of heron feeds on crickets, grasshoppers and similar small creatures. It also wades in water for fish and frogs. The cattle egret usually stays near herds of large grazers. Its harsh alarm calls and flapping escape help to warn them of predators.

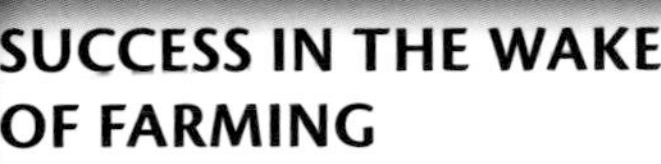

SUCCESS IN THE WAKE OF FARMING

Many animals suffer from the spread of people and farms across wild areas such as the African grasslands. But a few creatures benefit. The cattle egret is one of the bird world's success stories. In the wild it follows herds of gazelles, antelopes and other large plant-eaters. It pecks for the insects and other small creatures they disturb as they graze. As farm cattle, ploughs and other machines have spread, so have the egrets.

Avocet

The avocets seen in Europe in summer fly back to Africa for the winter. They feed on mudflats and estuaries, and also occasionally at inland lakes. (See also page 22.)

Egyptian goose

This goose lives not only in Egypt but across most of Africa and in parts of the Middle East. It can be a pest as it raids farm crops, especially the young soft plants. (See also page 18.)

Hippopotamus

Hippos must live in or near water. They rest, wallow and mate there. Also their skin is thin and quickly loses moisture. It needs a daily soak to stay healthy. (See also page 98.)

Great white egret

Almost twice the size of the cattle egret, this snow-white bird lives not only in Africa but in most other regions except parts of Europe. It eats almost any food.

If a Thomson's gazelle can sprint, swerve and keep ahead of a cheetah for more than 20 seconds, its chances of escape improve from 3 out of 10 to 9 out of 10.

The supreme hunters

- Cheetah
- Clouded leopard
- Jaguar
- Lion (male) and lioness (female)
- Snow leopard (ounce)

Big cats are the most fearsome hunters in the animal world. They stalk in stealthy silence, charge like lightning, pounce on prey with ferocious speed, stab the victim with their massive sharp teeth and slash with their cruel claws. The seven types of big cats are the tiger (see page 146), lion, cheetah, jaguar, leopard, snow leopard and clouded leopard. All are top predators in their habitats. However they sometimes come into conflict – for example, lions on the African plains sometimes kill and eat cheetah cubs. Cheetahs hunt mainly by day but the other big cats prowl mainly in twilight or at night. Lions live in groups called prides but all the others stalk alone.

Lion
The male lion's shaggy neck mane of thick fur makes him look even bigger and more ferocious. A typical pride of lions has 2–4 males. They patrol the pride's home area or territory, roar and leave urine scent-marks to keep out lions of other prides. They also defend the pride against threats such as hyaenas.

Clouded leopard
This is the smallest big cat, with a head and body about 80 cm long and a weight of 15–20 kg. It lives in thick forests from India across to Southeast Asia and spends most of its time in trees hunting birds, squirrels and monkeys. This leopard is such a good climber that it can run down trees head-first.

Cheetah
Famed as the fastest thing on four legs, the cheetah really does sprint at amazing speed. But usually for less than a minute since it has little endurance. Any longer and its prey, such as a Thomson's gazelle or impala, is likely to get away. Cheetahs live in dry, open areas in Africa, including the Sahara.

Lioness
Male lions rarely hunt. Catching food is the task of the lionesses, who are smaller and lack manes. They work as a team to surround and ambush prey such as zebras, antelopes and gazelles. Lions live in bush, scrub, grassland and woods across most of Africa. A few survive in the Gir Forest of north-west India.

Jaguar
The only big cat in South America, the jaguar likes water and swims very well. It creeps through swampy forests to hunt prey such as tapirs, peccaries, monkeys, turtles, fish, and rodents like capybaras. With the loss of forests due to farming and logging, some jaguars now live in rocky hills or dry scrubland.

Snow leopard
The long, thick fur of the snow leopard keeps it warm in the world's highest mountains – the Himalayas of Central Asia. Its paws are very wide and have fur between the sole pads to grip slippery ice. Snow leopards prey on a range of animals, from birds and rats to large wild cattle and sheep.

The cheetah can reach a top speed of about 100 km/h.
Its average chase when hunting is 150–200 m long and lasts about 20 seconds.

Close cousins, big and small

- Aardvark
- African elephant
- Asian elephant
- Rock hyrax

The two kinds of elephants are the largest land animals in the world. They also come from a very ancient group of mammals. Various kinds of elephants and mammoths have lived on Earth for more than 30 million years. The elephant's trunk, which is really its very long nose and upper lip, is like a multi-purpose fifth limb. It can grasp and pull leaves and similar food into the mouth for chewing. It sucks up water and squirts it into the mouth when the elephant drinks. The trunk also sniffs the air for scents. Elephants live in small herds, usually of females and young. The trunk is very important for touching, stroking and smelling other herd members.

Asian elephant

Slightly smaller than the African elephant, the Asian elephant also has smaller ears and shorter tusks. The tusks are very large upper teeth called incisors. They are made of a hard white substance, ivory. The elephant uses them to dig for food and water and to defend itself against enemies such as tigers.

African elephant

A large male stands more than 3 m tall, has a head and body nearly 7 m long and weighs over 5 tonnes. The tusks grow through life. Cows (females) have tusks too but they are much smaller. The massive ears help the elephant to hear tiny sounds and they also work like flapping fans to keep its bulky body cool. An elephant eats about 150 kg of food each day – the weight of two adult people. It includes grasses, leaves from trees and also twigs, bark and roots. An elephant herd is led by one or two experienced older females called matriarchs who know where to find food and water. Young males form groups known as bachelor herds. Old males or 'tuskers' usually live alone.

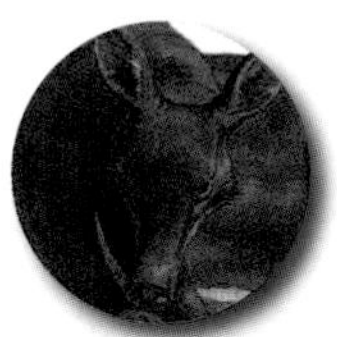

Aardvark

The name *aardvark* means 'earth pig' in the Swahili language of Africa. But this curious creature, about 2 m long from nose to tail, is not a type of pig. In fact it is not like any other mammal, although its closest cousins are probably the elephants. Like these giant relatives, the aardvark belongs to a very ancient group of mammals and it lives on the African savannah (grasslands). It comes out at night to search for its food of ants and termites, which it licks up with its long, sticky tongue. The aardvark has only a few weak teeth and does not chew. Its tiny prey are ground up in its strong, muscular stomach. Aardvarks travel 20 km or more each night for food. They live alone in deep burrows up to 15 m long.

Rock hyrax

Hyraxes (hyraces) look like large rats or small bears. But they are a separate group of mammals – and the closest living cousins of the elephants. Rock hyraxes eat grasses and similar plants. They live in groups in the drier parts of Africa, especially on inland cliffs and among the rocky outcrops known as kopjes.

An old bull (male) African elephant may be over 70 years of age and have tusks 3 m in length.

Tall and slim, big and fat

- Baringo giraffe
- Hippo (hippopotamus)
- Masai giraffe
- Nubian giraffe
- Okapi
- Pygmy hippo
- Reticulated giraffe
- Transvaal giraffe

The giraffe is one of the most peculiar mammals, with its extremely long neck and its ungainly, stick-like legs. Giraffes are easy to spot in their favoured habitat, the open bush of Africa, where they use their height to reach tasty twigs and shoots almost 7 m above the ground. There is only one main kind or species of giraffe, but this includes several varieties each with a distinctive coat pattern. The rare okapi of West African forests is a close cousin of the giraffe. Hippos are also hoofed mammals, like giraffes and okapis. They spend much of the day almost submerged in rivers or lakes, and come out at night to graze on nearby grasses and other plants.

Okapi

The okapi is such a shy forest-dweller that it is seldom seen. It gathers leaves using its curly dark-blue tongue which is almost 50 cm long!

Hippo

Hippos often lie in the water with just their ears, eyes and nostrils visible at the surface. Each hippo herd occupies a stretch or territory of river. Male hippos sometimes fight each other for the territory or for females at breeding time. They can inflict nasty wounds on each other with their long, tusk-like teeth.

Pygmy hippo

This mini-hippo is only about 90 cm tall but it is just as tubby as its big cousin and weighs up to 250 kg. It is found in the swampy forests of West Africa but is very rare, with just a few thousand remaining. Pygmy hippos feed mainly on grasses, roots and shoots which they find by grubbing on the forest floor.

Giraffe

Giraffes are by far the tallest animals. An adult male can grow to almost 6 m. Although the neck is so long it has only seven neck bones inside, just like other mammals (and ourselves). The giraffe's tongue is about 45 cm long, tough and flexible, and used to twist twigs and leaves from even the thorniest trees.

Distinctive giraffe varieties include the reticulated giraffe (shown opposite in the centre, just behind the okapi), with large chestnut patches separated by thin white lines. The Masai giraffe (upper left) has irregular, often star-shaped patches. Also shown are the Baringo giraffe (upper centre), two Transvaal giraffes (upper right and middle right) and a Nubian giraffe (lower right). Each giraffe's coat pattern stays the same through its life.

The male hippo is the third-biggest land mammal after the two types of elephants. It can be more than 4 m long and weigh well over 3 tonnes.

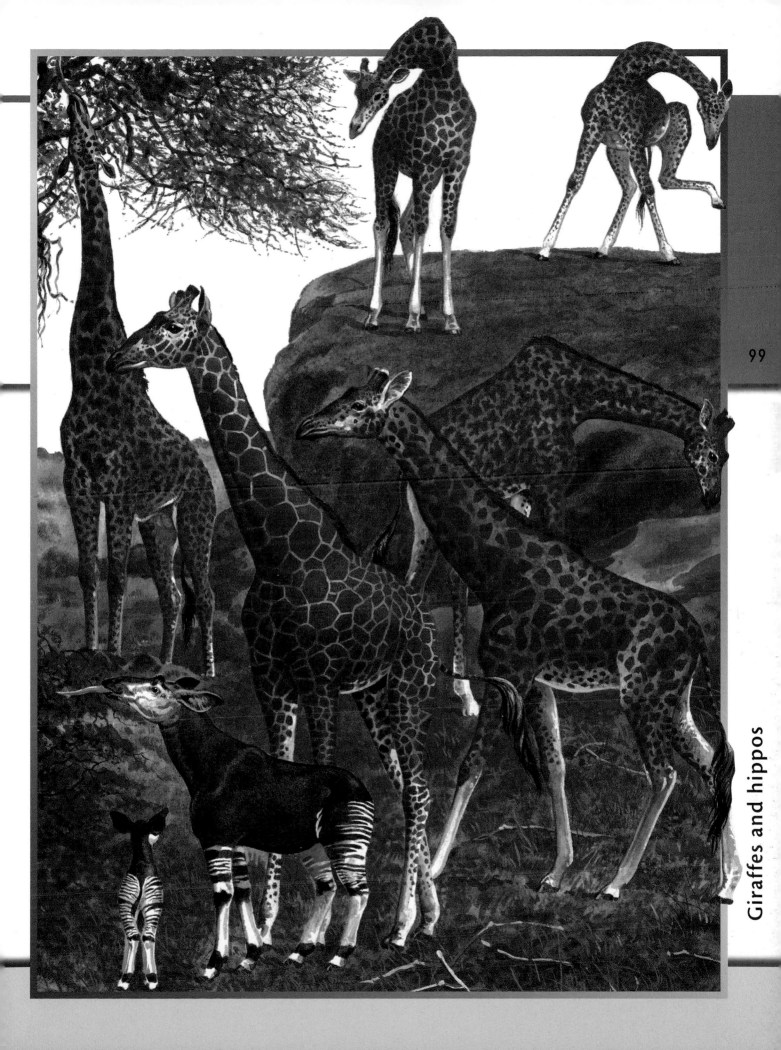

Making a living from dying

- American black vulture
- Californian condor
- Egyptian vulture
- Griffon vulture
- King vulture
- Lappet-faced vulture
- Turkey vulture

Vultures and condors are huge birds specialized to soar high in the sky, spot dead animals, swoop down, and peck at their bodies until only the skin and bones are left. (Sometimes not even these.) American vultures and condors form one group. Some detect food partly by its rotting smell. The vultures of Europe and Asia are a different group, more closely related to hawks and eagles. But all of these birds have long, broad wings and spend the daylight hours gliding over the land without flapping, carried high by winds and rising columns of air called thermals. They play one of nature's vital roles as recyclers, getting rid of the remains of other creatures.

American black vulture
Small for a vulture, with a wingspan of 1.5 m, this bird scavenges rubbish over much of South America and into southern North America.

King vulture
Unusually for a vulture, this type inhabits tropical forests in Central and South America. It uses smell rather than sight and can sniff out a dead animal hidden by the trees.

Griffon vulture
This is the commonest large vulture in southern Europe, usually seen soaring in flocks. It has a short, square tail and nests mainly on dry cliffs.

Californian condor
This magnificent condor is almost extinct. At one time there were fewer than 50 birds left in the world. Attempts have been made to rescue the species by hatching eggs and raising the chicks in captivity, and then returning the young birds to the wild. But its future still hangs by a thread.

Egyptian vulture
This small, pale vulture has long, narrow wings and a wedge-shaped tail. The adult is creamy white with black flight feathers.Young birds are dark brown but gradually become paler over four years. The Egyptian vulture is found from the Mediterranean region south into Africa and southern Asia.

Turkey vulture
The common turkey vulture does indeed have a turkey-like head. It ranges from southern Canada right through North and Central America to the tip of South America. It lives mainly in dry, open places including farmland. Known locally as the turkey buzzard, it eats any kind of food, even animal droppings.

Lappet-faced vulture
This is the biggest African vulture, with a wingspan of more than 2.5 m – almost as large as the black vulture of Asia. With side-swipes of its massive beak it tears lumps of flesh from recently killed animals on the African plains. Many vultures gather at lion kills after the big cats have eaten.

The world's largest bird of prey is the Andean condor.
It weighs about 10 kg and has a wingspan of some 3 m.

Thundering hooves on the plains

- African wild ass
- Asiatic wild ass
- Chapman's zebra
- Donkey
- Grevy's zebra
- Mountain zebra
- Plains zebra (common or Burchell's zebra)
- Przewalski's horse

If wild horses, asses and zebras were all the same colour, they would be very difficult to tell apart. They are all hoofed mammals, or ungulates, and close cousins of rhinos and tapirs. They are built for grazing on grasses and other low plants, and for galloping at high speed for long distances across open country. The rare Przewalski's horse of the Mongolian steppes (grasslands) is probably similar to the ancestor of today's many domestic horse breeds. Wild asses are smaller than most horses and have longer ears, less even manes and tufted tails. Zebras live mainly on the savannahs of East and Southern Africa.

Grevy's zebra

Grevy's zebras live mainly in north-east Africa, especially Ethiopia and Somalia, where the land is dry, rocky, scrubby and almost desert-like. Grevy's is the largest of the three kinds (species) of zebras, with a head and body about 2.8 m long and a weight of 400 kg. Its narrow vertical stripes are more black than white.

Asiatic wild ass

Mainly a desert dweller, this ass ranges from northern India and Tibet west to Iran and Syria. It is larger and more horse-like than the African wild ass and has a browner coat, but it lives in small family herds like its cousin. There are various local names for different varieties such as the onager and kiang.

Plains zebra

Zebras spend much of the day grazing, on the lookout for lions, hyaenas and African wild dogs. They live in small groups of a stallion (male) with 5–10 mares (females) and their foals (young).

Chapman's zebra

This is a darker variety of the plains zebra, with faint lines between the main stripes. Zebras gather in huge herds to find new grass.

African wild ass

Found in north-east Africa, this ass is very like its descendant, the donkey. There may be a dark shoulder line or faint zebra-type leg stripes.

Przewalski's horse

This wild horse was rescued from extinction by breeding in captivity. It stands only 1.3 m tall. Unlike a domesticated horse it has a stiff, upright mane.

Mountain zebra

The smallest and rarest zebra lives in the mountainous grasslands of south-west Africa. It has a slim build with narrow black stripes and a white belly.

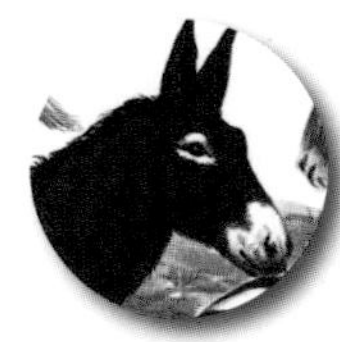

Donkey

The donkey is a domesticated form of the African wild ass. It is usually grey but some are brown, as shown in this breed called the Spanish giant donkey.

Przewalski's horses disappeared from their natural wild home of Central Asia in the 1960s. Hopefully captive-bred members released there will thrive again in the wild.

Huge eyes in the dark forest

- Golden potto (angwantibo)
- Lesser bushbaby
- Needle-clawed bushbaby
- Philippines tarsier
- Potto
- Slender loris
- Slow loris
- Thick-tailed bushbaby

They may look like big-eyed squirrels, and they have similar lives as they leap and bound through trees. But the bushbabies, pottos, lorises and tarsiers are active at night and they are in the primate group, related to lemurs, monkeys and apes. Like monkeys they have relatively large brains, well-developed hands with gripping thumbs, and large forward-facing eyes. Bushbabies live in Africa and catch insects using their sight and also their amazing hearing. Slower-moving are the lorises and pottos of Africa and Asia. They creep silently along branches as they sniff out their food of fruits, tree sap, insects, grubs and even the occasional small bird.

Thick-tailed bushbaby
This is the largest bushbaby with a nose-tail length of 75 cm. It lives in the forest and bush of East and Southern Africa and eats many foods including lizards and birds.

Needle-clawed bushbaby
This bushbaby's claws really are needle-sharp. Like other bushbabies it uses its long, fluffy tail to balance and turn in jumps.

Slow loris
Found in many parts of Southeast Asia, the slow loris 'freezes' when disturbed. It can remain completely still for an hour until its predator loses interest.

Slender loris
India and Sri Lanka are home to the slender loris. It eats many insects, even poisonous ones such as hairy caterpillars. It rubs off the stinging hairs before swallowing its prey.

Golden potto
This potto is a slow-moving creature of the night. It lives in the rainforests of Central Africa where it feeds mainly on maggots, caterpillars and similar grubs.

Potto
Resembling a teddy bear, the potto clambers along and underneath branches in the forests of Central Africa. It eats almost anything from sap to bats.

Philippines tarsier
Tarsiers are found only in Southeast Asia, including parts of the Philippines. They have huge eyes and very flexible necks, so the head can swivel almost all the way around like an owl. The tarsier sits still on its branch. When a lizard, mouse or insect ambles past it quickly pounces.

Lesser bushbaby
The commonest bushbaby, this type is found in many parts of Africa, mainly in drier woods and open bushland. Bushbabies keep in touch with each other by loud, twittering calls as they spring and leap through the trees at night. By day they sleep together, huddled in a tree hole or the crook of a branch.

Bushbabies are named after their child-like faces with large eyes, and also after their eerie wails at night which sound like a human baby crying.

Fast, elegant plains-dwellers

- Blue wildebeest (brindled gnu or brindled wildebeest)
- Bontebok (blesbok)
- Gemsbok (Beisa oryx)
- Gerenuk
- Klipspringer
- Pronghorn
- Sable
- Scimitar oryx (white oryx)
- Topi (sassaby)

Antelopes have keen sight, excellent hearing and sensitive noses, and they can run, dodge, swerve and jink at great speed. This is just as well, because these mainly grazing mammals are at great risk from predators such as lions, cheetahs, leopards and wild dogs. Many antelopes live on the African savannah (grassland), gathering in large herds for safety in numbers. They wander in search of the rains which bring fresh grass. Unlike the antlers of deer, which are shed each year, the horns of an antelope grow through life. Horn length and shape vary – in gazelles and small antelopes they are short, while the sable and oryx have long, curved horns.

Blue wildebeest
This cow-like antelope is strong and sturdy rather than slim and speedy. It forms huge herds on regular migrations across the African savannahs.

Topi
Topi are at home in greener, more marshy regions of the open savannah. They gather into herds of 3000 or more and travel to fresh grazing.

Klipspringer
These dainty antelopes live in stony, rocky places and jump with great agility. Their hooves are small and rubbery to give a firm grip on the rock.

Scimitar oryx
Named after its sword-shaped curving horns, this antelope was once widespread in North Africa. It now lives in a small area near the Red Sea.

Sable
The sable antelope has extremely long horns, measuring more than 1.5 m around their graceful curve. It prefers damp grassland and open woods.

Pronghorn
The only antelope in America, the pronghorn inhabits open prairie. It is one of the fastest runners in the world and, unusually, sheds its horns yearly.

Bontebok
The purple-brown bontebok prefers open grassland in Southern Africa. In the 1930s it almost become extinct but wildlife protection laws have helped to increase its numbers.

Gemsbok
The gemsbok is a type of oryx. It lives in dry savannahs across large regions of East and Southern Africa. It stands 1.2 m tall at the shoulder and its horns can be more than 100 cm long.

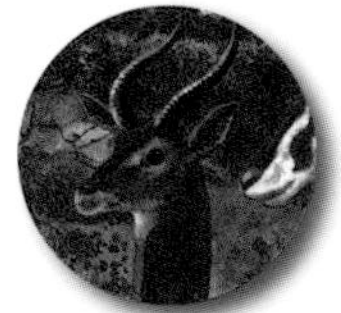

Gerenuk
The graceful gerenuk balances on its back legs and stretches its long neck upwards to eat the foliage of trees and tall bushes. Only the male gerenuk has horns.

The pronghorn can run at up to 90 km/h and sustain this speed for many minutes, unlike its rival for fastest land animal, the cheetah.

The nursery swamp

- African buffalo
- African jacana (African lilytrotter)
- African swamp rat
- Allen's swamp monkey
- Avocet
- Bearded reedling (bearded tit)
- Common waterbuck (Defassa waterbuck)
- Common garter snake
- Coypu (nutria)
- Dragonfly larva (nymph)
- Emperor dragonfly
- Great white egret
- Mink
- Painted terrapin
- Ram's-horn snail
- Sacred ibis
- Scarlet ibis
- Sitatunga (marsh nyala)
- Water snail

In and out of water

Swamps are found in most places. Largest is the Amazon Basin of South America.

Marshes, swamps, bogs, and the muddy overspill areas of lakes and rivers, are all types of the habitat known as freshwater wetland. These places are mosaics of small still pools and random-shaped islands, dotted with scattered clumps of reeds and rushes, the occasional thicket of bushes and trees or a few flowing channels. Wetlands are in-between worlds where both land and water are never far away. Most creatures which live here can walk, run, slither, swim and dive. The animals shown below raising their young come from swamps and marshes all around the world. But they have similar adaptations to this muddy, boggy, in-between habitat.

Great white egret

This large, white-feathered bird is so widespread that it has many local names, including great African egret and American egret. The female lays 2–5 eggs and both parents take turns to sit on, or incubate, them for 26 days. They also both feed the chicks when these hatch. (See also page 93.)

Common waterbuck

Waterbuck live in Africa in all kinds of habitats, from grassy plains to rocky hills to thick woods – but they are never far from water. If they are disturbed they race into the marsh and hide in the thick reedbeds. Few of their predators, which include lions and leopards, follow them into the soft mud or sinking sand. Waterbuck eat mainly young, soft plant shoots, especially grasses. They are strongly built antelopes standing about 1.4 m tall at the shoulder, with a weight of 200-plus kg. The male has sweptback horns with many rings along its length. He leads a small family group of females and their young. In times of drought, as the waterholes and pools shrink, waterbuck may come together into herds of 50 or more.

Common garter snake

Garter snakes are widespread across most parts of North America, apart from deserts and the driest scrub. They usually have three pale stripes, one along the back (upper surface) and one along each side, but apart from this, their colours and detailed patterns are very varied. Garter snakes grow to about 1.2 m long and hunt in marshes and damp undergrowth for small creatures such as salamanders, frogs, toads, fish, worms and insects. In the southern parts of their range they are active all year round. In the colder northern parts they come together in suitable caves or holes to hibernate through the winter. The female's eggs hatch while they are still inside her body and her young are born small but fully formed.

Mink

The American mink is very similar to the European mink, except it is usually slightly larger, and it lacks the white patch of fur on its upper lip just under its nose. Both species are expert swimmers and prowl wetlands for fish, frogs, waterbirds and crayfish. The female gives birth to 5–6 babies and feeds them for 2 months.

The female garter snake gives birth to as many as 80 babies, one of the highest number of offspring for any snake.

Allen's swamp monkey

Many monkeys avoid water, and some are terrified of it. But Allen's swamp monkeys splash about with no concern. They even search through the cloudy water and mud with their hands to grab prey such as water snails, fish, crabs and insect grubs. However their main food is fruits, soft nuts, young leaves and similar plant matter. These monkeys live in swampy areas of the tropical rainforests in West and Central Africa. They are strongly built with a head and body some 50 cm long, a tail of about the same length, and relatively short arms and legs compared to most monkeys. The baby clings to its mother for about 3 months.

African jacana

A widespread bird across nearly all of Africa, this jacana has a beak-tail length of 30 cm. It feeds on grubs, worms and soft shoots. Both male and female care for the eggs and young. Sometimes the mother carries her chicks on her back, in the manner of a swan.

Sacred ibis

The areas of black feathers and black skin on this ibis vary over its huge range, from Europe and Africa across South and Southeast Asia. In addition to the usual swamp diet of fish, frogs and snakes it may also kill and eat waterbirds.

Scarlet ibis

One of the world's most distinctive birds, the scarlet ibis is brilliant red all over except for its dark beak and a few dark flecks at the ends of its longest wing feathers. It is found in the north and north-eastern parts of South America, mainly along the coasts and in swamps and rainforests. The scarlet ibis roosts, nests and feeds in groups. It wades in shallow water and jabs its beak into the mud for fish, shellfish, worms, frogs and snakes. The female ibis lays two eggs and both parents take turns to sit on them for about 3 weeks. The chicks leave their nest at 5 weeks old.

Sitatunga

This marsh-dwelling antelope occurs in Central and Southwest Africa, especially along the Congo and Zambesi rivers. The male is slightly larger than the female with a head and body about 1.6 m long and a weight of 110 kg. He also has darker fur and his spiral-ringed horns (which the female lacks) can be up to 90 cm long. The female has one young or calf each year. Sitatungas usually take to the water to escape from predators such as lions and hyaenas. If the pool is deep enough they can hide by sinking almost entirely below the surface, with just the nostrils showing to breathe.

Bearded reedling

The male bearded reedling does not really have a 'beard' of dark face feathers, it's more a 'moustache' shape. These birds are found in reedy marshes across Europe and Asia. The 5–7 eggs hatch into fluffy chicks with gaping yellow- and red-rimmed mouths. Both parents feed them for 12 days.

THE PROBLEMS OF LIFE IN SOFT MUD

It is very easy to sink and get stuck in soft mud or 'quick sand'. So swamp animals move cautiously. They usually keep to clumps of plants where the roots make the mud firmer. Many creatures have big feet to spread their body weight over a large area and prevent sinking. The very long toes of the jacana (lilytrotter) mean it can walk over floating leaves such as water lilies. The sitatunga has long, wide, splayed hooves for the same reason.

Sacred ibises were worshipped in Ancient Egypt. They were kept as pets, and after they died some of them were preserved as mummies and buried with their owners.

Birds of shores and shallows

- Atlantic shag
- Blue-footed booby
- Brown pelican
- Dalmatian pelican
- Darter (anhinga or snakebird)
- Pied cormorant
- Red-billed tropicbird
- Red-footed booby

In addition to waders and gulls, many other types of birds can be seen fishing or foraging along the sea coast, or diving for food in the shallows. Cormorants and shags, with their dark plumage, are often spotted on shores, estuaries and harbours, and also on inland lakes well stocked with fish. Like their cousins the pelicans, they have stretchy, balloon-like throat pouches. The cormorant swims below the surface after its prey with amazing speed and agility. However its feathers become waterlogged. This helps the bird to stay under the water. But it also means that, after its meal, the cormorant must sit for a long time with its wings held out to dry.

Darter

Sometimes called the snakebird because of its long and bendy neck, the darter is an expert fish-catcher. It lives on rivers and lakes and uses its sharp, dagger-like beak to spear fish underwater. The darter swims low in the water, often with just the head and neck visible – which also makes it look like a snake.

Brown pelican

Unlike other pelicans, the brown pelican feeds from the air. It soars about 15 m high over the water, then folds its wings and plunge-dives to catch a fish in its chin pouch. Brown pelicans are found along the coasts of North and South America and also around the Caribbean and Galapagos Islands.

Pied cormorant

Like most cormorants, the pied cormorant of Australia and New Zealand nests in large colonies. It sometimes gathers in flocks of thousands.

Atlantic shag

Looking like a small cormorant, the shag has beautiful, glossy, dark green feathers. It prefers rocky coasts and breeds in small colonies on cliff ledges.

Red-billed tropicbird

Rocky ledges on tropical islands are the main breeding sites of these birds. They spend most of their time gliding over the open sea, rarely landing except to breed.

Dalmatian pelican

Some pelicans are fairly common but the Dalmatian pelican is on the official list of threatened species. Only about 2000 survive on lakes in eastern Europe and across Russia to China. The pelican likes shallow water – but so do people, for fishing and boating. They easily disturb this timid bird.

Red- and blue-footed boobies

Boobies are open-ocean birds that plunge-dive for fish, like their close relatives the gannets. The red-footed booby is common over tropical oceans. Unlike its cousins it nests in trees rather than on ledges or the ground. Blue-footed boobies breed on the Galapagos Islands of the East Pacific.

Cormorants are sometimes used by people to catch fish. The owner keeps the bird on a string tied to a ring around its neck. The ring stops the cormorant swallowing its catch.

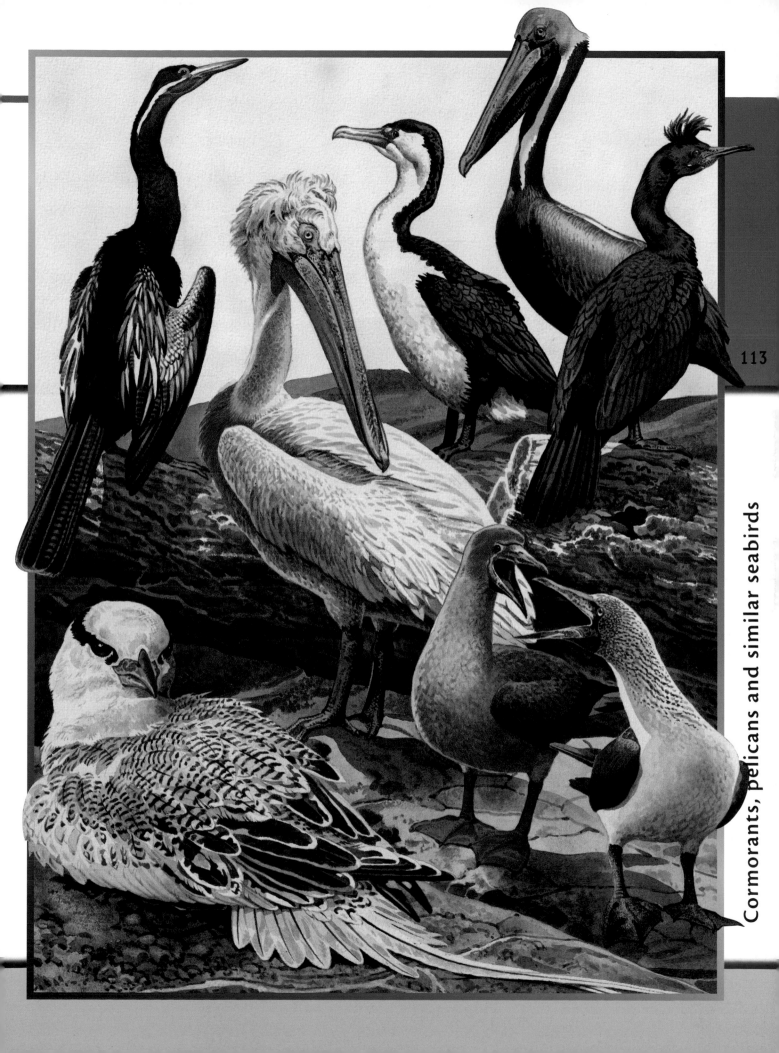

Hard-shelled slow-movers

- Alligator snapping turtle
- Painted terrapin (pond turtle or painted turtle)
- Spike-shelled turtle
- Spiny softshell (spiny terrapin)
- Spotted turtle

Turtles and terrapins are far from fast. But they are well protected from harm inside their hard body shells. These mainly water-dwelling creatures, along with the land-living tortoises, are members of the reptile group called chelonians. There are about 180 different kinds of turtles and terrapins living in rivers, lakes, swamps and other freshwater habitats, mostly in northern and tropical regions of the world. Many eat a mixture of plant parts, especially soft leaves and stems of water weeds, and small animals such as worms, pond snails, shellfish, baby fish and young frogs. Female turtles lay eggs on land, in sand or mud or hidden under stones and logs.

Painted terrapin

Yellow stripes along the head and neck, reddish lines on the legs and bright markings around the 14-cm-long shell identify this common North American turtle. However its patterns are very varied, especially on the bright yellow underside. It eats mainly long, trailing water plants and also water grubs.

Spotted turtle

Most turtles are not active hunters. They lie in wait for passing victims or chomp leisurely on plants. Their dull colours and mottled patterns, like the spotted turtle's dotted patches of yellow, orange or red, help to camouflage them among the shady weeds and stones at the bottom of a lake or river.

Spike-shelled turtle

Young, newly-hatched turtles have softer shells than the adults. They are at risk from many predators such as herons, fish-eagles, mink and large fish. The spikes around the edge of this turtle's shell gradually lengthen and harden over the first two years for excellent protection.

Spiny softshell

With a shell up to 45 cm long, the spiny softshell is a powerful predator of fish, crayfish, water insects and even small water birds. Its name comes from the small, spiky lumps on the front of the shell above the neck. This turtle lives in quiet ponds and creeks in eastern, central and south-eastern North America.

Alligator snapping turtle

This is the largest freshwater turtle in North America, growing to more than 75 cm long and 90 kg in weight. It lies on the bottom of a muddy lake or slow river, its ridged shell camouflaged by weedy growths to look like a jumble of stones. The turtle holds its mouth wide open to reveal a small, narrow, pale, fleshy flap on the floor of its mouth. This wriggles like a worm and attracts fish, crayfish and similar animals. If they come to check the 'bait' the turtle snaps shut its massive, sharp-edged jaws and swallows the victim whole – or slices it in half. As an alternative, the turtle can lunge upwards and grab passing prey in the sharp, hooked front parts of its jaws. The female alligator snapper lays 20–40 eggs in early summer.

Snapping turtles feed on carrion such as drowned deer and pigs, finding them by smell. They have been used to locate the bodies of people murdered and thrown into deep lakes.

The widest, toothiest grin

- American alligator
- Black caiman (black alligator)
- Nile crocodile
- Gharial (gavial)

Crocodiles have been lurking in rivers and swamps since the time of the dinosaurs, more than 150 million years ago. The crocodilian group has 22 members including 14 species of crocs and seven types of alligators and caimans mainly from Central and South America. The final member is the curious gharial from the Indian region with its very long, slim, tooth-studded snout. It catches fish by a sideways sweep of its jaws.

Gharial
This is one of the most aquatic crocs, rarely leaving the water except to breed. It has more than 100 small, pointed teeth for grasping slippery fish prey.

Nile crocodile
The Nile croc lives in many watery areas of Africa. It grabs large animals or birds which come to drink, drags them under the surface to drown, then tears off chunks to swallow.

American alligator
Once rare in the wild, this alligator has recovered its numbers and lives across the south-east USA. The mother 'gator' builds a nest mound of old plants and lays her eggs inside. The plants rot and release heat which incubates the eggs for about nine weeks. Like many croc mothers she guards the eggs and watches over the babies for several months.

Black caiman
Caimans are similar to alligators and live mainly in South America. Largest is the black caiman of the Amazon region at 4.5 m long. It was hunted so much for its meat and leathery skin that it is now excessively rare.

The largest living reptile is the saltwater crocodile, also called the Indo-Pacific, Australian or estuarine crocodile. It grows to more than 7 m in length.

Standing tall in the water

- Bald ibis
- Boat-billed heron
- Greater flamingo
- Hammerkop
- Maguari stork
- Royal spoonbill
- Yellow-billed stork

The 60 different kinds (species) of herons are found around the world. Most are grey or brown but some smaller types, known as egrets, are pure white. The storks, with about 17 species, are similar to herons – tall birds with long legs and long beaks. Herons prefer marshes and lakes while most storks stride across grassland and scrub. All of these birds feed on small animals which they catch on the ground. They are slow but graceful in the air. Herons tuck in their long necks when flying, storks hold their necks stretched out straight. Spoonbills and ibises are also tall birds that mainly wade in water, feeling for prey with their sensitive beaks.

Greater flamingo
The greater flamingo stands up to 1.2 m tall and is one of the strangest birds with its spindly legs, long neck and peculiar bent beak. The beak works as a sieve to filter tiny animals like shrimps from the water, as the flamingo swishes it from side to side in a shallow lagoon.

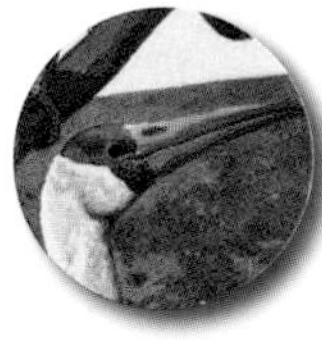

Yellow-billed stork
A common waterbird in Africa, especially in eastern parts, the yellow-billed stork lives on lakes, marshes and sometimes on coasts. Like certain other storks and also the bald ibis, it has bare skin around its face. Otherwise its face feathers would get wet and dirty while feeding.

Boat-billed heron
Most herons have long, thin, pointed beaks. The boat-billed heron has a broad, flattened beak like the hull (body) of a ship. This heron is not quick enough to stab fish or frogs. Instead it scoops its broad beak through the water to catch little shrimps, grubs and other small, slow-moving animals.

Hammerkop
The hammerkop from Africa is a strange relative of the storks. Its name means 'hammerhead' because its thick beak and the feathers on the back of its head make it look like the end of a hammer. Its nest is also unusual – a huge, untidy dome up to 2 m high. The nest may take almost six weeks to construct.

Maguari stork
The Maguari stork lives on the pampas – the wide, open grasslands of Argentina and nearby countries in South America. It eats mainly insects.

Bald ibis
This ibis lives in dry scrub and nests on cliffs. It has become rare in recent years. This may be due to its habitat becoming even drier, partly due to water being pumped away for farmland.

Royal spoonbill
Found in Australia, New Zealand and New Guinea, royal spoonbills build big stick nests in tall trees near water. They wade in the shallows to catch small prey.

One huge member of the stork group, the marabou stork of Africa, has the longest wings of almost any land bird. They are more than 3 m from tip to tip.

Gills, lungs and fins

- African lungfish
- Australian lungfish
- South American lungfish

Lungfish look like links with the distant past, when prehistoric fish first crawled from the water, developed fins into limbs, gulped air and began to live on land as amphibians. Lungfish belong to a very ancient fish group which has been around for more than 300 million years. But they have continued to change or evolve through time and are now well adapted to life in slow muddy rivers, weedy lakes and shallow swamps. Like other fish, they can breathe oxygen dissolved in the water using their gills. But if the water lacks oxygen, for example when it is very shallow and warm, lungfish can also swallow air into their tube-like lungs.

African lungfish

This eel-like fish grows to 2 m long and is a fearsome hunter of smaller water creatures such as fish, frogs, crayfish, lizards and water birds. At breeding time the male wriggles and digs a hole in the sand or mud for the female to lay her eggs. He guards these while they develop and hatch.

Australian lungfish

The Australian lungfish is from a different ancient fish group to the South American and African types. It has fins with strong, fleshy bases like the famous 'living fossil' fish called the coelacanth. Also its Y-shaped, two-lobed lung is higher in the body, above the main guts, compared to the other lungfish. And it cannot survive buried in the mud if its creek or pool dries out. But like its cousins, the Australian lungfish is a powerful predator of almost any small water animal. It reaches a length of 1.5 m. These lungfish live naturally in the Mary and Burnett Rivers of north-east Australia. They have also been taken to other waterways in the region, in case accidental pollution or some other problem threatens this unique fish.

WHEN LUNGFISH GO TO SLEEP

South American and African lungfish can survive drought, when their water dries up, by burrowing into the damp mud beneath. As the dry season arrives and the rivers and pools shrink, the lungfish noses and presses the mud aside to form a vase- or tube-shaped chamber. It curls up in here and then its skin makes a layer of mucus (slime). This goes hard to form a waterproof lining or cocoon for the chamber. If the water disappears completely the lungfish seals the top of its chamber with another lump of mucus. Here it can pass the dry season, its body processes working very slowly, almost like a mammal in hibernation. This method of survival in the hot, dry season is called aestivation.

South American lungfish

Like other lungfish, the South American type does not breathe air through its nostrils as in land animals. It gulps air down its throat and gullet (oesophagus) and through a slit into its lungs. This lungfish reaches about 1.2 m in length and, like its African relative, its lower fins are long feelers.

In its cocoon in the mud, a lungfish can survive for several years. It does not eat, but gets its nutrients and energy by breaking down its own muscles.

Bursting with life

- Anaconda
- Blue-fronted parrot (blue-fronted Amazon)
- Blue-headed parrot (blue-cheeked Amazon)
- Blue-legged tree frog
- Blue morpho butterfly
- Cuvier's toucan
- Double-collared aracari
- Emerald tree boa
- Fork-tailed wood nymph
- Jaguar
- Military macaw (great green macaw)
- Orange cock-of-the-rock
- Orange leafwing (Anaea ryphea) butterfly
- Ornate hawk-eagle
- Popelaire's thornbill hummingbird
- Rare tiger (Heliconius ethillus) butterfly
- Red-legged honeycreeper (purple-legged honeycreeper)
- Scarlet macaw
- Squirrel monkey
- Sulphur-breasted toucan
- Toco toucan
- Woolly spider monkey

The greatest wilderness

Northern South America has the largest tropical rainforests on Earth.

124

No one knows how many kinds, or species, of animals live in the world's great tropical rainforests. Certainly these warm, green, humid places have more wildlife than all other habitats on Earth added together. Greatest is the Amazon rainforest, a vast area of swamps, low-lying hills and jungles surrounding the world's biggest river, the Amazon. This huge region stretches across central and northern South America, from the Andes mountains in the west to the Atlantic Ocean in the east. Each new area explored by wildlife experts reveals yet more spectacular animals and fascinating plants which are new to science.

Military macaw
There are 18 kinds or species of macaws in the parrot group. The military macaw is one of the larger members, about 75 cm in total length. It has a red forehead which looks like a soldier's peaked cap. This macaw is found in forests from Mexico through Central America and south to Brazil.

Scarlet macaw
One of the largest members of the parrot family, the scarlet macaw, measures nearly 90 cm from beak to tail-tip. Once a female and male court and mate they stay as a pair for many years, flying and roosting and feeding together. They eat all kinds of plant food including very hard seeds and nuts.

Jaguar
The jaguar is a medium-sized big cat with a head-body length of about 1.4–1.6 m. In many ways it is the South American version of the leopard in Africa. It has a similar powerful and muscular build and rosette-type spots, climbs and swims well, and is adapted to various habitats. (See also page 94.)

Orange cock-of-the-rock
No other bird has such a bright orange colour or the double fan of feathers on the head, as this 30-cm-long rainforest dweller. But only the male is so colourful. The female is dull brown for camouflage. She does the work of making the nest, sitting on the eggs and then feeding the chicks.

Orange leafwing butterfly
The upper surfaces of this butterfly's wings are brilliantly coloured with glowing orange and pink. But when it lands and closes its wings above its back, only the lower surfaces can be seen. These are marked with streaky greens and browns to look exactly like the leaves all around.

Emerald tree boa
Most boas are constrictors. They squeeze victims to death in their coils. But the emerald tree boa grabs prey in its mouth, which is equipped with large, strong fangs. It grows to just over 100 cm in length and its bright green colour, flecked with white and yellow, make it look like a mossy branch.

The emerald tree boa strikes so rapidly and accurately that it can snatch fast-flying birds and bats as they swoop past.

Blue morpho butterfly

The male of this morpho type of butterfly has the most intense, shining blue colour of almost any animal. It measures up to 15 cm across its wing tips and flits around the tops of trees.

Fork-tailed wood nymph

This hummingbird has a total length of 10 cm, and one quarter of this is its long beak. It hovers in front of flowers, probing inside to reach nectar. Its richly coloured feathers have a metallic sheen.

Cuvier's toucan

The huge and colourful beak of the toucan is made mainly of a lightweight, spongy, horny substance. It is designed both for eating fruits and for displaying to attract a partner at breeding time.

Squirrel monkey

This small, agile monkey leaps and bounds through the trees like its namesake, the squirrel. It is about the same size as a squirrel too, with a head and body 30 cm long and a tail of about 40 cm. It feeds mainly on fruits and a variety of small animals.

Anaconda

This massive and muscular snake grows to about 9 m long. It is a member of the boa and python family and kills prey by wrapping itself around the victim and gradually tightening its coils. As the victim breathes out the anaconda increases its grip so the victim cannot breathe in, and it soon suffocates. Anacondas lurk in swampy forests and feed on wild pigs, deer, tapirs, peccaries, large fish and even alligators or caimans. Like other boas the female anaconda does not lay eggs but gives birth to baby snakes, each 60 cm long.

Rare tiger butterfly

The yellow-orange with black stripes gives this butterfly its name. However like many tropical butterflies, it is found in different colours such as bright yellow-green. Its wingspan is 8 cm.

Toco toucan

This is the best-known of the 38 kinds (species) of toucans. It also has the largest beak, both in actual size at 20 cm long, and in proportion to its body. The beak has a sharp, slightly wavy or serrated edge for chopping and cutting up fruits and berries.

Spider monkey

No other monkeys can swing and leap through the trees as well as spider monkeys. Their hands have long, strong fingers that hook over branches. However the thumb is tiny so the spider monkey cannot grab or hold food easily in its hand. But it can do this with its tail, which is very strong and curling or prehensile. The underside of the tail tip has no fur and the skin is ridged like our fingertips for a good grip. The spider monkey can hang by one hand, pick fruits and flowers and other foods with its tail, and pass these straight to its mouth.

RIOTS OF COLOURS

Why are some animals so brightly coloured or patterned that predators can see them from far away? There are usually two reasons. One is to attract a partner at breeding time. In the struggle for life it is important not only to survive, but also to breed. Bright wings or plumage which are more likely to impress a mate, outweigh the risk of being spotted by a predator. The second reason is a warning to predators: 'I taste horrible!' Combinations of certain colours, in particular red and black or yellow and black, warn that a creature is poisonous or has a foul taste.

The anaconda is not quite the world's longest snake (the reticulated python is). But it is the bulkiest or heaviest snake, weighing up to 200 kg.

Prehistoric 'pigs' of the forest

- Baird's tapir
- Brazilian (South American or lowland) tapir – adult
- Brazilian tapir – young
- Malaysian tapir – adult
- Malaysian tapir – young
- Mountain (woolly or Andean) tapir

Tapirs are something of a leftover from the prehistoric age. Fossil remains of their bones and teeth show that the tapirs roaming forests more than 25 million years ago were hardly different from today's versions. These creatures resemble pigs but they are close cousins of rhinos. They eat plants and live mainly in woods and forests. There are four kinds or species, all shown here. Largest is Baird's tapir, but only just. A typical tapir has a head and body about 2 m long, a small tail of up to 10 cm, and stands around 100 cm tall at the shoulder. It is a strong, sturdy, thick-set animal that weighs 250 kg or more – as much as four adult people!

Brazilian tapir
This is the widest ranging type, found across the northern half of South America. It is also the only tapir that lives in grassy plains as well as forests. Like the others it feeds at night and eats all kinds of plant food such as leaves, grasses, stems, buds, shoots, flowers and fruits. Sometimes it raids corn and other farm crops.

Baird's tapir
This tapir has a neck mane of stiff, bristly hairs. It is found from Mexico down to northern South America. Like other tapirs in the region its main enemy is South America's big cat – the jaguar. If a tapir is attacked it bites hard, kicks even harder, thrashes about and then crashes off through the thick undergrowth.

Malaysian tapir
This is the only tapir not from South America. It's from – Malaysia! Also Thailand and other parts of Southeast Asia. Its striking black and white pattern probably helps with camouflage in its dense rainforest home. It breaks up its body outline and recognizable pig-like shape. All tapirs are excellent swimmers.

Young Brazilian tapir
Tapirs live alone except for a mother with her baby, or when male and female come together for a day or two to mate. This happens at any time of year. The baby is born 13 months later in a nest or den deep in the forest. It hides here, staying perfectly still, while the mother goes off to feed.

Mountain tapir
Forests up to 4500 m high on the massive Andes Mountains are home to this tapir. It is well protected against the night cold by its long, woolly coat. Despite its clumsy appearance, it can easily climb steep slopes and scramble over loose rocks. A tapir's eyesight is not so keen and it finds food mainly by smell.

Young Malaysian tapir
Very different from its parents, this baby tapir has a red-brown coat with white spots and stripes. The pattern helps to conceal it in the forest's dappled shade. After two months the pattern begins to fade. By six months old the youngster is black and white and almost ready to leave its mother.

Tapirs have four toes on each front foot but only three on each back foot.

Hard-shelled swimmers

- ▸ Green turtle
- ▸ Hawksbill turtle
- ▸ Leatherback (leatherback turtle)
- ▸ Loggerhead (loggerhead turtle)

Large tropical rivers like the Amazon are sometimes visited by a very specialized group of reptiles – marine turtles. These slow, heavy creatures can hardly move on land. The females haul themselves ashore every two or three years to lay their eggs, usually burying them in beach sand. Then they return to the water, where they spend most of their lives wandering in search of food. The youngsters hatch in two to three months, dig themselves up to the surface and scurry to the waves, avoiding predators like gulls, lizards and otters. In water the turtle swims elegantly with its large, flipper-like front legs. These flap rather than row, almost like flying underwater.

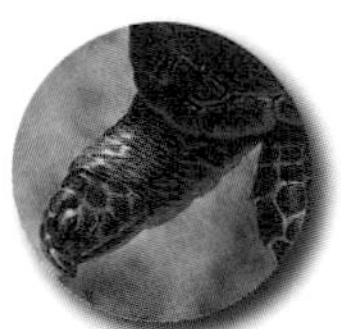

Loggerhead
The loggerhead is named after its large, bulky head which contains very powerful jaws. Turtles have no teeth but their jaws have very sharp, horny edges. The loggerhead can easily crunch up crabs, prawns, whelks, oysters and similar shellfish, as well as sponges and seaweeds. It grows to about 100 cm long.

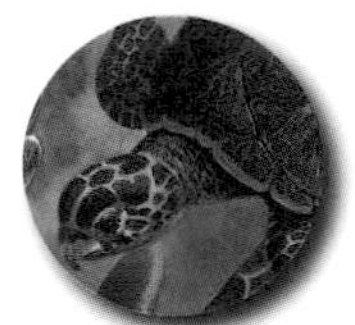

Hawksbill turtle
The shell of this turtle has beautiful swirling patterns and was used as the original 'tortoiseshell'. Like other large turtles, the hawksbill is now protected by wildlife laws. It is named after its narrow, hooked mouth, which it pushes into rocky crevices for prey such as crabs and fish.

Green turtle
Seaweeds, and especially seagrasses, are the green turtle's main food. But, like other big turtles, if it is hungry it eats whatever may be available, including jellyfish and shrimp-like krill. Green turtles grow to about 1.2 m long and were once widely hunted for their shells, meat, eggs and to make 'turtle soup'.

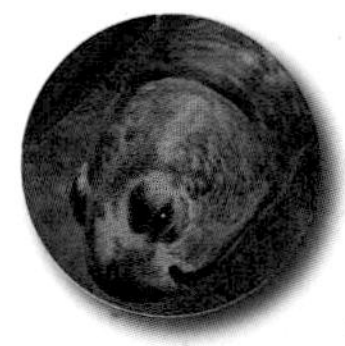

Leatherback
The massive leatherback is the largest living turtle. Like other turtles and tortoises it has a two-part body casing. The hard domed part over its back is the carapace. The flatter part on the underside is the plastron. In a typical turtle these are made of flat slabs of bone covered with plates of a hard, horny substance. But the leatherback lacks the bony slabs and horny plates. Instead it has a casing of tough, leathery skin, almost like the rubber of a car tyre. Leatherbacks have weak mouths and feast mainly on jellyfish.

The leatherback turtle grows to 1.7 m long and weighs 600 kg. And its huge front flippers can measure 2.6 m across.

Tiny, brilliant masters of the air

- Common swift
- Crimson topaz
- Frilled coquette
- Popelaire's thornbill
- Ruby-throated hummingbird
- Sword-billed hummingbird

The Amazon region teems with tiny, brilliantly coloured birds that dart among the leaves like feathered jewels. They flap so fast, their wings look like a blur and sound like a low droning buzz or hum. These are the hummingbirds. Most can dash forward at great speed yet stop on a dot to hover in mid air, and then fly backwards or even straight up like a miniature rocket. No other birds are so speedy and aerobatic – except perhaps for their close cousins, the swifts. The common swift spends more time in the air than almost any bird except the great albatross. It feeds, courts, mates, rests and even sleeps on the wing.

Sword-billed hummingbird
Most hummingbirds have long beaks to probe into flowers, and long tube-shaped tongues to sip and suck the nectar. The sword-billed has the longest beak compared to its body size of any bird. It not only reaches into the deepest funnel-shaped flowers but also pecks up small insects hiding there.

Crimson topaz
In the Amazon rainforest a crimson topaz darts across a sunlit clearing. Its two black tail feathers trail behind. It's a male impressing his mate. The female lacks tail streamers but she too has glittering plumage. She differs from most female hummingbirds, which are duller than males for camouflage when nesting.

Popelaire's thornbill
Thornbill hummingbirds live mainly in the cooler forests on the slopes of the Andes Mountains. They sip up nectar and also snap up small grubs, beetles and similar creatures. The speed of flapping is average for a hummingbird at about 50 wing beats per second.

Frilled coquette
Normally shy and hidden by leaves, frilled coquettes forget their secretive habits at breeding time. Female and male resemble butterflies as they flutter and dance in mid air. Like all hummingbirds they feed mainly on sweet, sugary flower nectar. Only this has enough energy to power their fast-flying lifestyle.

Ruby-throated hummingbird
Only the male has the glowing throat colour – the female is white in this area. Ruby-throats are among the world's smallest migrating birds. On wings only 12 cm across they fly from summer breeding areas in eastern North America almost 1000 km to winter in Mexico, the Caribbean and Central America.

Common swift
The swift twitters and flits around trees, rocks, riverbanks and buildings as it catches gnats, midges and similar tiny flying insects. Swifts winter in Africa, then fly north and east to Europe and Asia to breed. Their feet are so small and weak that they can hardly perch on twigs, only cling to cliffs and walls.

Hummingbirds flap their wings faster than any other birds, more than 80 beats per second in some types.

Long-tongued ant-lickers

- Burmeister's armadillo
- Fairy armadillo
- Giant anteater
- Naked-tailed armadillo
- Nine-banded armadillo (common long-nosed armadillo)
- Pichi
- Silky anteater (two-toed anteater)
- Small-scaled tree pangolin
- Tamandua (tree or collared anteater)

Ants and termites fight and bite when caught. But they are so tiny that big ant-eating mammals such as anteaters, pangolins and armadillos have little to fear. They lick and pick up their miniature victims dozens at a time with their long sticky tongues.

Tamandua

Tamanduas are smaller, tree-living versions of the giant anteater. They can use the strong, flexible, furless tail to wrap around branches like a fifth foot. The tamandua can walk on the ground but it is slow, clumsy and at greater risk from predators. These anteaters live in the forests of Central and South America.

Giant anteater

The giant anteater is a very odd-looking mammal with its brush-like tail, shaggy fur and long, curved nose. It grows to 2 m long including the tail and shuffles through wood or scrub in Central and South America. It digs open a nest using its very long, sharp claws, laps up a few hundred insects and moves on.

Small-scaled tree pangolin

This is one of four African pangolins. The long, curving, prehensile (grasping) tail helps it to climb in the branches, using its sharp-edged scales to get an even better grip on the bark. Tree pangolins hunt at night for nests of ants and termites high in trees. They sleep in their own tree nests by day.

Silky anteater

This anteater has soft, golden fur like a cuddly toy. But it can slash out with its fearsome sharp, curved front claws. Also, like other anteaters, it does not destroy a nest while feeding. It makes a small hole, takes some occupants, then leaves the ants to multiply and repair their nest, so that they can be a future meal.

Nine-banded armadillo

Armadillos eat mainly ants, termites and other small creatures such as woodlice and worms. But some types, especially the nine-banded armadillo (opposite, shown lower right) are also fond of fruit. There are 20 species of armadillos living in South and Central America, with a few north to Mexico and the southern USA. They have very keen noses to detect food in the soil and strong front claws to dig rapidly for roots, termites and ants. Smallest is the fairy armadillo from southern South America (lower left) at only 15 cm long. It has less armour-plating and spends much time burrowing in sandy soil. Also shown are the naked-tailed armadillo (centre left), pichi (centre right) and the small Burmeister's armadillo (lower middle).

When feeding the giant anteater flicks its 60-cm-long tongue in and out twice every second.

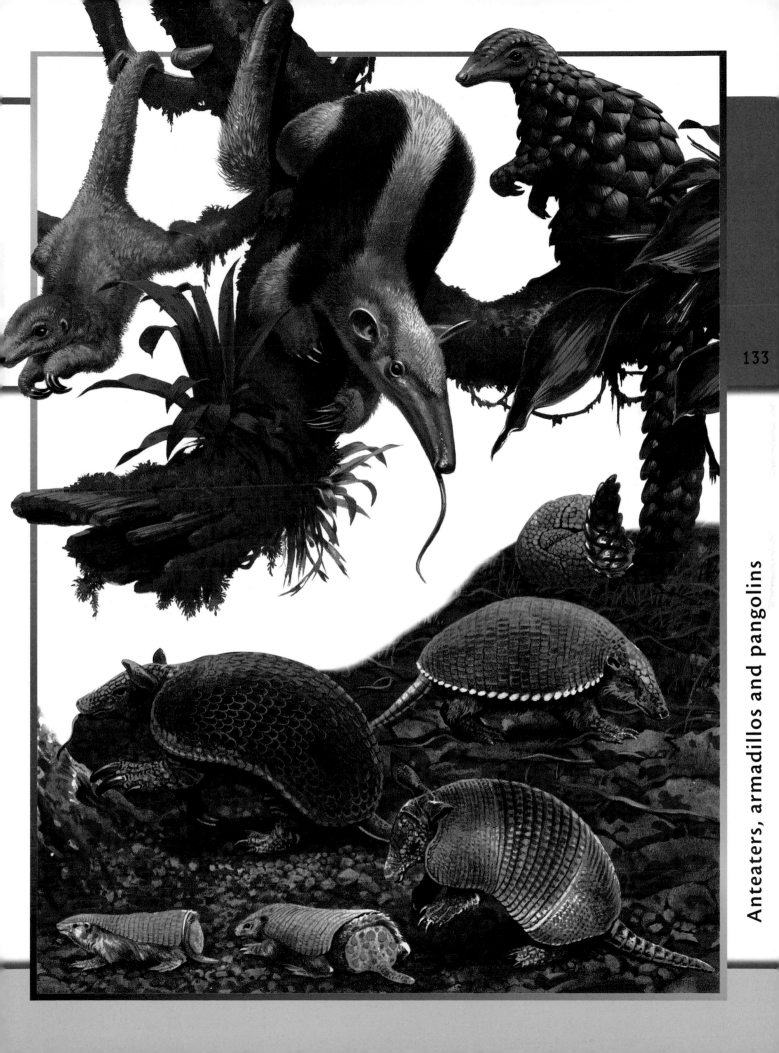

Fish of the stagnant swamps

- Arapaima (pirarucu)
- Aruana (arowana)

The bony-tongue fish are exactly as their name suggests. The tongue in the floor of the mouth is strengthened by hard, plate-like pieces of bone. When the fish bites it does not so much bring its two jaws together, as press its tongue up against the roof of its mouth. Both the tongue and mouth roof have tooth-like projections and these crush and grind the prey – mainly other fish. Bony-tongues hunt in tropical lakes, rivers and swamps. They are mostly large and powerful fish, able to burst at speed from a weedy hiding place to ambush prey. The best-known member of the group is the arapaima – one of the biggest freshwater fish in the world.

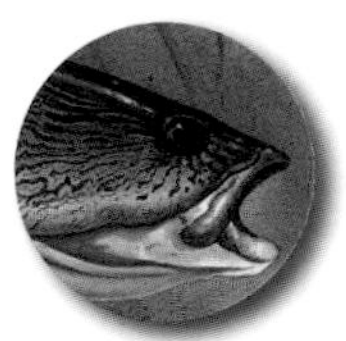

Arapaima

The arapaima is known by various local names in its swampy home of tropical South America. It can 'breathe' in the normal fish way using its gills and also by gulping air down its throat into its swim bladder. The swim bladder's blood-rich lining then works like a lung to absorb the oxygen from the air into the body. This is very useful since warm, still, stagnant swamp water is very low in oxygen.

Tales were told of arapaimas swallowing full-grown people. This is very unlikely but the arapaima is still enormous for a freshwater fish. Its vast mouth can swallow a 50-cm prey whole. At breeding time the arapaima scoops a nesting hollow in the sandy swamp bed and guards its eggs there. It also guards the babies when they hatch.

Aruana

The aruana has a very distinctive shape, its flat-topped head making a straight line with its back and almost running into its tail. The fish grows to about 100 cm long and inhabits lakes and slow-flowing rivers in warmer parts of South America. The fleshy 'tentacles' on its upturned mouth are barbels. These are very sensitive to touch. They can also feel water currents and detect or taste certain substances in the water. Like the arapaima, the aruana has a long dorsal (back) fin and a similar anal (underside) fin, both set near the rear of its body just in front of the tail. When the fish swishes its rear body powerfully these fins help the tail to thrust it forward very quickly, usually to dash at prey.

MONSTER FISH FROM THE STEAMY JUNGLE

Many tales came from the remote swamps of the Amazon region about giant fish. Some were said to be so huge that they could swallow a person, an anaconda (the world's bulkiest snake) or a caiman (South American crocodile) in one gulp – or even a whole canoe! However it is difficult to judge the length of a fish half-hidden in muddy water. Also the size of the fish probably grew each time the story was told. The arapaima can swallow large prey but it feeds mainly on smaller food including shellfish and worms. Several other freshwater fish probably grow as large, including the massive catfish of rivers such as the Mekong in Southeast Asia.

There are sightings of giant arapaimas growing to more than 5 m in length and 200 kg in weight. But 3 m and 100 kg are probably more realistic.

Birds with babysitters

- Common coucal (crow pheasant)
- Didric cuckoo
- Great spotted cuckoo
- Hoatzin
- Purple-crested turaco
- Red-billed (Renauld's) cuckoo
- Senegal (chestnut-winged) coucal

The 127 members of the cuckoo family are spread all over the world. But not all of them sing 'cuck-ooo!' and lay their eggs in other birds' nests. Most cuckoos in Europe, Africa and Asia do – they are brood parasites. The female places each of her eggs in the nest of another pair of birds. The new parents do not notice the extra egg and so become unknowing 'babysitters', feeding and raising the hungry chick. Across the ocean in the Americas, most cuckoos build their own nests and care for their own chicks in the usual way. Coucals are types of cuckoos that live mainly in Africa and Asia, on the ground or in the low bushes of scrubby countryside.

Hoatzin One of the world's strangest birds, the hoatzin dwells in the Amazon's densest swamps. It is about the size of a chicken, eats only plant food and flies weakly, preferring to clamber through the branches. The hoatzin chick has finger-claws on its wings to help it climb – a feature found in no other living bird.

Purple-crested turaco Turacos are close relatives of cuckoos but live only in Africa. They eat almost entirely fruits, including berries that would quickly poison most animals (and people). Other birds give their growing young at least some meat in the shape of grubs, slugs and worms. Yet turaco chicks are fed mostly fruits.

Great spotted cuckoo This cuckoo lays its eggs in other birds' nests, but the chick is not quite as violent as in other cuckoos. It does not tip out the unhatched eggs and chicks of the parent bird. But it does eat most of the food that they bring. Also, being so big and strong, it may lean on the other chicks and squash them to death.

Red-billed cuckoo The red-billed cuckoo of Southeast Asian rainforests stays mainly on the ground and is strong enough to tackle mice and lizards. Like most cuckoos it can fly and run well but is not a great climber. It has the typically powerful, partly hooked cuckoo beak for grabbing all kinds of food.

Senegal coucal Locusts, beetles, mice, frogs and the chicks of other birds are all tasty meals for this cuckoo. Its odd call sounds like water bubbling from a spring.

Common coucal Common indeed, this adaptable coucal can survive on a wide variety of food in forests, grassland and farmland. Its call is 'boob-boob-boob'.

Didric cuckoo Instead of 'cuck-ooo' this African member of the family sings 'dee-derr-rik', which has become its first name. It clings to tree trunks and pecks at the bark for grubs and caterpillars.

If a pair of small birds like warblers raise a cuckoo chick, the 'baby' grows to be 10 times the size of the parents!

Slower than a snail

- Three-toed sloth (brown-throated three-toed sloth)
- Two-toed sloth (Linné's two-toed sloth)

Few animals can match the sloth for slowness – not even a snail. A sloth can easily sleep for five days, stay awake but not move for 12 hours, and take three days to move along a 3-m branch. There are five kinds or species of sloths and they are found only in the forests of Central and northern South America plus the nearby islands. They are so adapted to hanging in trees by their hook-like claws, that they cannot walk properly on the ground and tend to drag themselves along by their claws. Yet in some patches of forest they are by far the most common medium-sized mammals and seem to have very few predators. So slowness has had little effect on their success.

Three-toed sloth

A better name for this animal would be the three-clawed sloth since each set of fingers and toes ends in large, hard, sharp claws up to 10 cm long. The sloth uses these as branch hooks and also to slash at the occasional enemy such as an eagle or jaguar. It feeds by day and night on buds and soft young leaves, hanging quite still for camouflage when at rest. The three-toed sloth is slightly smaller than the two-toed type, with a head and body 60 cm long, a short stumpy tail and a weight of around 4 kg.

Two-toed sloth

If fingers are on the front limbs, and toes are on the rear limbs, then this sloth is really the two-fingered sloth. Confusingly it has two fingers on each hand, but three toes on each foot just like the three-toed sloth. It eats mainly leaves with some soft shoots, twigs and fruits and is active (as much as a sloth can be) mainly at night. After a full feeding session, which lasts about 7 hours, almost one third of its body weight is the food inside its stomach! Sloths produce droppings only about once each week.

Mother sloth with baby

Sloths do everything slowly – even breed and grow. The baby three-toed sloth takes 6 months to develop in the womb of its mother before it is born. This is a long time for a mammal of its size. The baby two-toed sloth takes even longer, just over 11 months. Only one baby is born at a time and this clings to its mother, feeding on her milk, for about a month. Then it begins to crawl about on its own and lick up the part-digested leaves she brings up or regurgitates. Sloths can live for more than 25 years.

HANGING AROUND

The sloth has many strange body features for its unusual upside-down life. The joints in its arms, hands, legs and feet can be 'locked' into their bent positions so the animal is able to hang from a branch with almost no effort. It also has a lower normal body temperature than almost any other mammal, down to 30°C. (Our own temperature is average for a mammal at 37°C.) Also it's fur slants the 'wrong way' compared to other mammals, up towards its back, so rain runs off it easily.

A big meal of leaves may stay in the stomach of a sloth, being slowly digested, for almost one month.

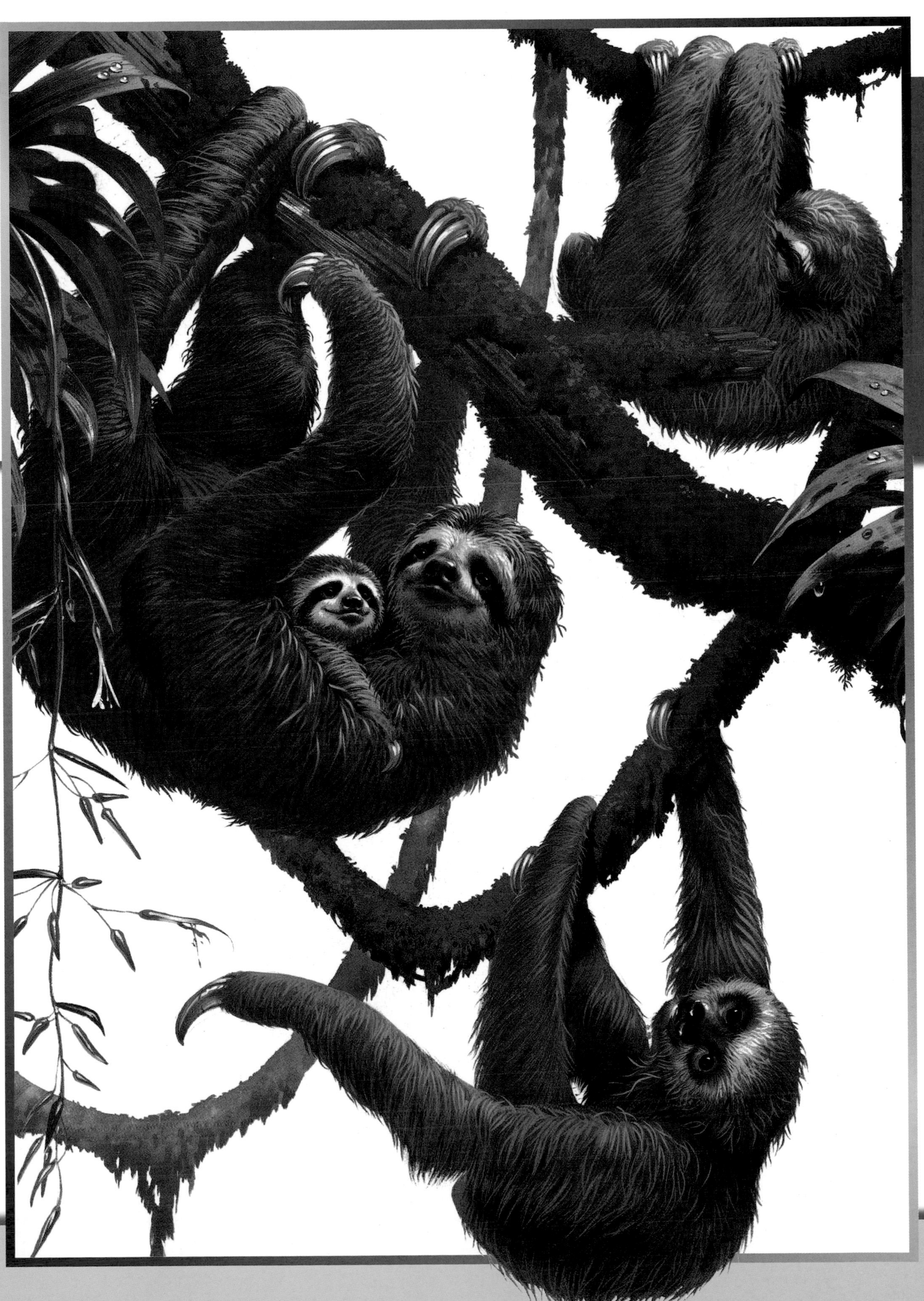

Gnawers of the Americas

- Capybara
- Cavy (wild guinea pig)
- Chinchilla
- Coypu
- Crested porcupine
- Hutia
- North American porcupine
- South American porcupine

Rodents vary from pygmy voles and pocket mice to rats, squirrels and beavers. Most bigger rodents live in the Americas, like the largest of them all, the capybara. The key feature of a rodent is its teeth. It has four big incisor teeth at the very front of the mouth, two in the upper jaw and two in the lower. They are shaped like chisels or spades with wide, sharp, straight ends. They never stop growing and if the owner never gnawed they would eventually pierce the inside of its mouth. However they are continually worn down as the creature bites, gnashes, nips and nibbles at tough plant food such as roots, twigs, bark, seeds and nuts.

Cavy
This is the original 'guinea pig' from which our many kinds of pet guinea pigs are bred. It lives in grassy and rocky scrub in southern South America.

Coypu
Coypus swim well and dig long riverbank burrows with their webbed feet. Originally from southern South America, they have escaped from fur farms in other regions such as Europe.

North American porcupine
Porcupines move with a slow, shuffling waddle. They need not run from enemies since they are protected by long, sharp spines.

Hutia
Hutias are rare rodents of the West Indies. They eat plants and also small animals like snails and lizards, climbing after them through the trees.

Crested porcupine
This porcupine has extra-big spines or quills up to 30 cm long on its back. Like most other porcupines it eats fruits, leaves, buds and similar plant food.

South American porcupine
Gripping the branches with its curly grasping tail and sharp-clawed toes, this porcupine is quite at home in trees.

Chinchilla
Chinchillas live on the rocky slopes of the southern Andes Mountains. Fast and agile, they leap on their strong back legs as they look for food to hold in their front paws and nibble. At night they sleep in groups of 50 or more, in holes among the rocks. Their long, soft fur and bushy tails keep them warm.

Capybara
Almost as big as a small pony, the capybara is never far from water. If a jaguar comes near the capybara can stay under the surface for five minutes – but then it might be grabbed by a crocodile! These huge rodents live in groups of 10–40 in the Amazon region. They feed on all kinds of grasses and water plants.

Porcupines cannot shoot out their quills as some tales suggest. But the quills have barbed tips and work their way deep into an attacker's body.

The jungle branches

- Black eagle
- Black-headed oriole
- Blue triangle butterfly (common bluebottle or Graphium sarpedon butterfly)
- Ceylon blue magpie
- Common eggfly (great eggfly or Hypolimnas bolina butterfly)
- Lion-tailed macaque
- Long-nosed tree snake
- Malabar pied hornbill
- Papilio polymnestor swallowtail
- Ring-necked parrot
- Silvered langur
- Slender loris
- Tickell's blue flycatcher
- Tokay gecko
- White-headed starling
- Yellow-fronted barbet

Tropical tree-dwellers

The Indian sub-continent has many unique creatures in its tropical forests.

Tropical rainforests like those in Sri Lanka and southern India are not only the richest and most varied places on Earth for wildlife, they are also the places where cycles of nature happen fastest. As soon as a piece of fruit falls or an animal dies, it is attacked by an army of worms, grubs, maggots, millipedes, mites, moulds and other recyclers. They quickly decay it and rot it into the soil. As it is broken down into raw minerals and nutrients, plants take these from the soil for their own new growth. All the raw materials for life are constantly in use. Such fast recycling means that tropical forest soils are surprisingly thin and poor – and little use for farming.

Malabar pied hornbill

This large hornbill is about 75 cm from beak-tip to tail-end. It has a very big beak too, with a lumpy extension on top called the casque. This hornbill takes a wide range of prey from grasshoppers and beetles to lizards and baby birds, plus plant food such as fruits (especially figs) and seeds.

Blue triangle butterfly

Also called bluebottles, these common butterflies vary greatly in colour across their huge range, from India through Southeast Asia to Australia. They come into gardens and parks to sip nectar from plants such as the 'butterfly bush', buddleia. They have wingspans of about 8–9 cm.

Black eagle

Forest birds of prey tend to have wings which are shorter from tip to tip, but broader from front to back, compared to birds of open plains. This shape gives better control in the air for sudden twists and turns among the trees. The black eagle snatches small monkeys, lizards, birds and other prey from the branches.

Common eggfly

This butterfly has gained from the spread of towns, villages and parks because its caterpillars feed on a wide range of garden plants. It has a wingspan of up to 11 cm. Like many tropical butterflies it has a number of varieties, called morphs, which differ markedly in colour from region to region.

Slender loris

The lorises are cousins of bushbabies and tarsiers. The slender loris, which has a head-body length of 20–23 cm, sleeps by day in a tree hole or the fork of a branch. As dusk falls it wakes up to feed. The loris sees well in the dark with its enormous eyes, as it steals slowly and silently along the branches, gripping securely with its long fingers and toes. Its big toe is opposable, like its thumb (and our own thumb), so it can wrap its foot around a twig. The loris creeps cautiously near a victim, locates it by sight and smell, and then suddenly shoots out its hands to grab the food. It eats many small animals such as grubs and caterpillars and it may also chew young leaves, soft shoots and buds. (See also page 104.)

The slender loris eats a wide range of poisonous or foul-tasting prey such as caterpillars, beetles, bugs, ants and millipedes.

Black-headed oriole
Several kinds of black-headed orioles live across Africa and southern Asia. They eat mainly soft-bodied insects like maggots and caterpillars, thereby helping farmers. But they also form flocks to feed on crops. They measure 20 cm from beak to tail.

Silvered langur
There are many different types of this familiar langur (also called the common or Hanuman langur) across the India region. The silvered variety has a brighter grey-white colour to its fur. It eats mainly plant foods and often raids crops and even shops for food.

Tokay gecko
This is a large lizard for a gecko, up to 30 cm long. It is a common and welcome visitor to houses, scampering over floors and up walls to catch flies, cockroaches, mice and other pests. Both parts of the name come from the male's mating call: 'Toh-keh, geh-koh.'

Ceylon blue magpie
This large and colourful bird, measuring about 45 cm from beak to tail, is familiar in parks and gardens on the island of Sri Lanka (formerly Ceylon). It is an adaptable feeder, taking all kinds of plant and animal matter.

Long-nosed tree snake
The back-fanged or rear-fanged snakes are a group of poisonous snakes which have their long teeth, or fangs, towards the rear of the upper jaw. (Vipers and other venomous snakes have their fangs at the front of the mouth). They include the boomslang, mangrove snake, vine snakes – and the long-nosed tree snake. Coloured green for camouflage among the leaves, it lies with the rear part of its body on a branch but the front part sticking out like a branch. It eats mainly lizards such as geckos and is an extremely fast mover through the trees.

Papilio polymnestor swallowtail
This type of swallowtail often flutters about garden flowers, feeding on their sugary nectar. It is a large, powerful and direct flier with wings measuring up to 13 cm from tip to tip. It lives only in Sri Lanka and southern India.

Yellow-fronted barbet
Also known as the golden-fronted barbet, this bird lives in many parts of India and mainland Southeast Asia. Like other barbets it has a large, strong, sharp beak for its body size. It pecks apart fruits and flowers – barbets are messy eaters.

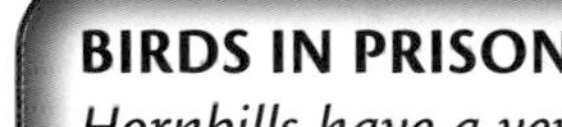

BIRDS IN PRISON
Hornbills have a very unusual method of nesting. After the parents mate, the female enters a hole in a tree trunk and stays there while the male builds a 'wall' across the entrance. He dabs on bits of earth and mud to leave just a small slit. He feeds her through this opening while she sits on the eggs. She cannot get out – but predators who might steal the eggs cannot get in. In the Malabar pied hornbill the female breaks her way out when the chicks are a few days old.

Lion-tailed macaque
Macaques are strong, heavy monkeys that are at home on the ground as well as in trees. The lion-tailed macaque of southern India not only has a lion-like tuft of hairs at the tip of its tail, but also a very showy ruff or mane of long fur. It eats a wide variety of food including fruits and insects.

The lion-tailed macaque is one of 11 species of macaque monkeys which are included in the official 'Red List' of threatened animals. It is classed as endangered.

Lords of the jungle

- Caspian tiger
- Indian tiger
- Siberian tiger
- Sumatran tiger
- White Indian tiger

Biggest of big cats, the tiger has inspired awe, respect and fear for thousands of years. It is a solitary, secretive, stealthy hunter of forest, scrub, long grass and tangled undergrowth known in India as jangle, *a term which was the origin of the word 'jungle'. All tigers belong to the same species,* Panthera tigris. *There are different varieties of this species across Asia. The Caspian tiger dwells to the east of the Caspian Sea, the Indian tiger in India and Bangladesh, the Indo-Chinese tiger on the mainland of Southeast Asia, and further varieties on large islands such as Sumatra and Java. Most massive and powerful is the great Siberian tiger of far East Asia.*

Caspian tiger

The most westerly variety, the Caspian tiger was found from Iran across southern Russia to western China. It is now very, very rare. Tigers hunt mainly large prey including various deer like sambar, chital, swamp and red deer, also wild pigs, gaur (wild cattle), water buffalo and even young rhino or elephants.

Sumatran tiger

One of the smaller varieties, the Sumatran tiger is about 2.5 m long from nose to tail and weighs up to 150 kg. It has a 'face ruff' of long cheek fur. Tigers are solitary – they usually live and hunt alone. If two tigers are together they are either a male and female at breeding time or a mother with her youngster.

Indian tiger

The Indian tiger is the most numerous variety yet it still numbers just a few thousand in the wild. Its main stronghold is the vast Sunderbans delta of mangrove swamps at the mouth of the River Ganges, on the India-Bangladesh border. Unlike many wild cats, tigers love water. They could lie in it all day – and do!

White Indian tiger

Most mammals produce very pale or very dark forms, which are born naturally from time to time of normal-coloured parents. The white tiger is a famous example and popular with zoos. It occurs mainly in north-east India and seems to survive just as well in the wild as its yellow-brown cousins.

Siberian tiger

The long, thick coat of the biggest tiger variety protects against the icy winds and thick snow of far eastern Russia and China. The Siberian tiger grows to more than 3.2 m from nose to tail. It is also more muscular and sturdy than other varieties, weighing over 320 kg. There are probably less than 200 left.

TIGER CONSERVATION

Rarely a tiger is killed because it keeps attacking cattle or humans. But tigers are also killed illegally for their body parts. For example, ground-up tiger bones are supposed to have mysterious powers to heal sick people. Medical tests show that there are no such powers. But still the poachers kill tigers.

A tiger is not especially brave against other animals. It is sometimes driven from its kill by a hungry hyaena or even a determined jackal.

The chatterbox birds

- Blue-fronted parrot
- Crimson rosella
- Eclectus parrot
- Fischer's lovebird
- Golden-shouldered parrot
- Grey parrot
- Hawk-headed parrot
- King parrot
- Masked lovebird
- Princess parrot
- Ring-necked lorikeet
- Slate-headed parakeet
- Sun conure
- White-breasted parrot
- White-fronted caique

There are more than 330 different kinds of parrots – a bird family which includes parakeets, lovebirds, macaws and cockatoos. They live mainly in the tropical regions of Africa, South America, Australia, Southeast and South Asia. Lories and lorikeets are colourful members from Southeast Asia and Australia. They differ from other parrots mainly in their feeding habits. They use their brush-tipped tongues to gather pollen and nectar from flowers. Conures, parrotlets and caiques are also in the parrot group. Some parrots are at risk from collection for the pet bird trade. But a bigger threat is destruction of their tropical forest homes. *(See also page 166.)*

Hawk-headed parrot
This South American type has a large hawk-like beak, a hawk-like ruff of feathers on its neck, and also hawk-like flight – fast, shallow flapping and gliding.

Fischer's lovebird
These birds live in grassland and woods in East Africa. Lovebirds are so called because the female and male often sit close together preening each other.

Grey parrot
Greys live in the forests of Central Africa, where they form communal roosts of up to 10,000 birds. Sadly, collecting for the pet bird trade has greatly affected their numbers.

Sun conure
Named after its bright orange-yellow plumage, the sun conure lives in open woodland in South America. It nests inside holes in tree trunks.

Princess parrot
A rare Australian type, the princess parrot likes rivers or lakes with plenty of eucalyptus (gum) trees along the banks. The male has a very long tail.

Blue-fronted parrot
Also called the blue-fronted Amazon, this bird lives mainly in South America. It feeds in groups in the treetops.

Crimson rosella
Common in many parts of Australia, this rosella often visits gardens and bird tables. When it flies, the blue wings stand out from the red head and body.

King parrot
This large, powerful Australian parrot prefers rainforest and damp woodland. Like most members of the family it feeds mainly on fruits and seeds.

Eclectus parrot
Female and male look like different species, so unlike are their colours. The female is red with a blue chest and shoulders, the male mainly green.

Some tame parrots have been taught to identify more than 50 different objects – especially when food is the reward!

Shy, secretive and sightless

- Amazon river dolphin (bouto or boutu)
- Ganges river dolphin (Ganges susu or side-swimming dolphin)
- Indus river dolphin (Indus susu)

Most dolphins live in the open sea. Their secretive and little-known relatives are the river dolphins from some of the largest rivers in the world. There are five kinds or species – two in the Indian region, two in South America and one in China. River dolphins have tiny eyes and are almost blind, because sight is little use in their muddy water. However they can swim fast and accurately using the squeaks and clicks of echolocation or sonar, like other dolphins (and also bats). They find their prey of fish and similar animals by sound too, grabbing the victims in their long beaks equipped with more than 100 small, sharp teeth.

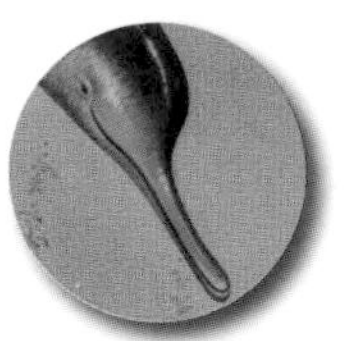

Amazon river dolphin

This freshwater dolphin lives in the Amazon, Orinoco and connected large rivers in South America. Although most river dolphins are quite shy, the Amazon dolphin is inquisitive and sometimes approaches a small boat, or even comes near a swimmer if she or he keeps still and quiet. These dolphins often move around in small, close-knit groups. They are very agile and frequently swim on their sides or even upside down. Amazon river dolphins from the Amazon itself tend to be lighter pink in colour and larger, about 2.3 m long and 120 kg in weight, compared to those from the Orinoco. Another dolphin, the tucuxi, also swims in the fresh waters of the Amazon but it is greyer and much smaller, only about 1.4 m in length.

Ganges river dolphin

This is one of the largest river dolphins, reaching a length of 2.4–2.6 m. It is a powerful swimmer and often leaps straight up like a missile bursting out of the water. The alternative name of 'susu' comes from the sound these dolphins make when they breathe out through their nostrils – the blowhole on the forehead. They eat many types of fish including carp and catfish, as well and shrimps.

Indus river dolphin

The Indus dolphin is very similar to the Ganges dolphin at about 90 kg weight. It is hunted for its body oil which is used in some local medicines. Like other river dolphins, its search for food has been severely affected by dams built for hydro-electricity and to water farm crops. All river dolphins are rare and affected by overfishing, pollution, drowning in nets and traps, and noisy boats which interfere with their delicate sonar systems.

The Indus river dolphin and the whitefin river dolphin of the Chiang Jiang (Yangtze) in China are two of the world's rarest mammals. Each numbers just a few hundred.

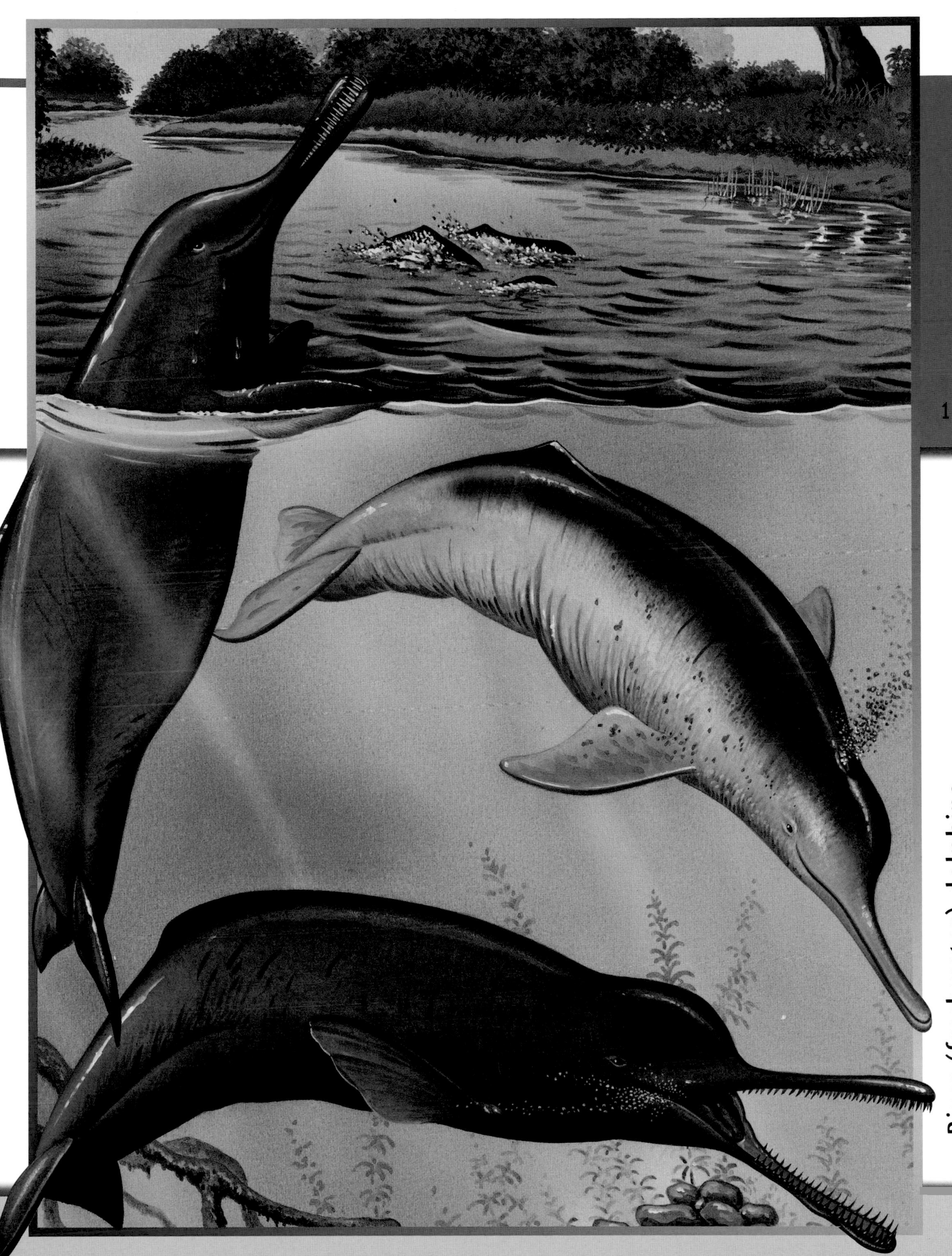

Monkeys with dog-like faces

- Black-and-white colobus (Angolan black colobus or white-epauletted colobus)
- Mandrill
- Moustached monkey
- Patas monkey
- Proboscis monkey

Monkeys are busy, active, agile, inquisitive creatures who live in groups, leap about in trees and make various whooping and screeching noises. The monkeys of Africa and Asia (Old World) do not have the grasping, prehensile tails of their American (New World) cousins. They also have longer muzzles with the nostrils close together so the face resembles a dog. But it can be a very colourful face, with fur or bare skin of glowing red, blue or yellow. The facial colours of full-grown adults usually become even brighter in the breeding season, showing that they are ready to mate. The backside or rump is often just as colourful, and for the same reason.

Moustached monkey

Like most monkeys the moustached monkey of West Africa is an 'opportunistic' feeder. This means it isn't a fussy eater. It consumes plant foods such as seeds, nuts, fruits, leaves, flowers and shoots. It also takes animal foods such as insects, grubs, lizards, birds' eggs and chicks – and anything else!

Black-and-white colobus

Familiar across the middle of Africa, black-and-white colobuses rarely come down to the ground. In the morning and evening they search for food in the tree tops. In the midday heat they relax and groom each other. The physical contact of grooming is important in keeping the group together.

Proboscis monkey

This rare monkey is named after its enormous nose, which is much bigger in the male. It lives only in the coastal mangrove swamps and riverbank trees of Borneo in Southeast Asia. It is a very specialist feeder, munching mainly on the leaves and fruits of mangrove and pedada trees.

Mandrill

This type of baboon lives in West African forests and, with its close relative the drill, is the biggest kind of monkey. A male mandrill has a head and body 90 cm long and weighs 50 kg. When it bares its fang-like canine teeth it is more than a match for leopards and other predators.

Patas monkey

Slim and long-legged, this African monkey can climb much better than a human and also run faster, reaching 50 km per hour! An old, experienced male leads his troop of about 10 females and their young. They feed by day on the ground, munching almost anything edible, and sleep at night in trees.

A MEAL WITH A FRIEND

Almost all kinds of monkeys live in groups, usually led by a mature, powerful male. Screeches and other calls are important to keep in contact. Grooming is too, and it also helps health. Troop members delicately pick fleas, lice and other pests from each other's fur and skin – and eat them.

In a well-fed proboscis monkey, one-third of the entire body weight is the stomach with the leaves inside.

A-maze-ing air-breathers

- Climbing perch
- Fighting-fish (Siamese or Thai fighting-fish, blue and red varieties)
- Three-spot gourami

The fish called anabantoids include many types familiar from the tropical aquarium – gouramis, combtails, paradise fish, climbing perches and fighting-fish. They are also called labyrinth fish because of the way they breathe. Normally they use their gills like other fish. But often the warm, still water of their tropical swamps and pools is low in oxygen. So the fish gulps air into two special chambers, one on either side of the head behind the eye. Each chamber has flaps that form a maze or labyrinth with a blood-rich lining. The air is trapped and its oxygen passes into the blood.

Climbing perch

About 25 cm long, the climbing perch lives in India, Southeast and East Asia. It makes good use of its air-absorbing labyrinth organs when it crawls from the water and wriggles across land, even over tree trunks and rocks. The perch usually does this to look for new water because its pool has almost dried up. It moves on land using its strong lower front fins (pectorals) and also the spikes on its gill covers, assisted by pushing with its tail. It can travel several hundred metres, usually at night when the air is cooler and damper, and there are fewer predators.

Fighting-fish

These small fish, only 5–7 cm long, live naturally in ponds and sluggish rivers in the Thailand region. The males vary from green to brown while the females are lighter olive-brown. However they have been bred for thousands of years to produce many varieties of fighting-fish, in different colours and sizes. Some have very long fins which they spread out as a threat to the rival.

Three-spot gourami

Wild gouramis live mainly in Southeast and East Asia, some reaching about 60 cm long. But many types have been selected and bred by people to produce a wide range of aquarium fish. These include thick-lipped, kissing, dwarf, lace, honey, sparkling and croaking gouramis. They are usually peaceful but at breeding times the males may attack other fish.

WHY FIGHTING-FISH FIGHT

Fighting-fish do not fight for fun or because they dislike each other. It is a natural instinct of many male animals at breeding time to show that they are strong and fit, and so a suitable mate for the female. Also fish may contest a living area or territory, where they feed and which they also need to possess in order to attract a mate. In the wild most such 'fights' normally involve body postures and displays rather than actual physical battle. However even in the wild the male fighting-fish is quite aggressive. It lifts its gill covers, stretches out its fins and may actually attack other male rivals. People have made these instincts stronger by selecting the most agressive males for breeding.

Fighting-fish can become so aroused that they even attack their own reflections on the inside of the aquarium's mirror-like glass side.

Plump, game and colourful

- Argus (argos) pheasant
- Golden pheasant
- Indian peafowl (peacock)
- Mikado pheasant
- Swinhoe's pheasant
- Turkey

Gamebirds are mostly large, plump-bodied and spend much time on the ground. They are known as 'game' because they were hunted for their tasty meat, and still are in some regions. The group includes pheasants, grouse, partridges, snowcocks, francolins, quails, turkeys, guinea fowl and guans. The pheasants in particular have amazing plumage – especially the males (cocks) at breeding time. Their long, flowing feathers are beautifully patterned in dazzling shades of blue, red, green and gold to attract a mate. The females (hens) tend to have drab brown plumage for camouflage. It is their job to build the nest and raise the chicks. *(See also page 180.)*

Argus pheasant The male argus pheasant of Southeast Asia has very long, ornamental wing feathers to impress his possible breeding partners. He prepares a special arena in a forest clearing by moving leaves and plants. Then he calls to the females before beginning his leaping, dancing courtship display.

Swinhoe's pheasant A rare species, Swinhoe's pheasant is found in the wild only on the island of Taiwan, like the mikado pheasant. This gamebird is small and delicate compared to other pheasants. It has suffered through loss of its habitat in recent years, as its tropical forest home is destroyed.

Mikado pheasant This pheasant is very rare and now protected by law on its home island of Taiwan in East Asia. A special reserve has been created to protect both it and Swinhoe's pheasant. In many areas pheasant eggs and chicks are eaten by introduced animals such as cats and rats.

Indian peafowl The massive tail fan of the male or peacock is one of nature's most spectacular sights. It is covered in eye-like spots and the huge feathers quiver and rattle during his courtship display for the female or peahen. The peafowl's home was India and nearby countries but it has now spread around the world.

Turkey The wild turkey of North America lives in forest and scrub, where it feeds on the ground eating seeds, nuts and berries. At night it flaps into the branches to rest. The various kinds of farmyard turkeys, popular food at events such as Christmas and Thanksgiving, have been bred from this wild relative.

Golden pheasant Like many of its pheasant relatives, the golden pheasant is often kept in bird collections for the male's beautiful plumage. This gamebird came originally from the forests of China. During courtship the cock spreads the fan-like feathers at the sides of his head to cover his face and beak.

The peacock never, ever lays any eggs. This is because he is the male bird. It's the female, or peahen, which lays eggs.

A jungle clearing

- Anoa (dwarf buffalo)
- Babirusa (Sulawesi forest hog)
- Banded lisang
- Blue-trimmed crow butterfly (Euploea eleusina)
- Celebes macaque (crested macaque monkey)
- Crested mynah
- Dwarf cuscus (Sulawesi cuscus)
- Fiery-browed starling
- Flying dragon (Wallace's flying lizard)
- Muller's parrot (blue-backed parrot)
- Paris peacock butterfly (Papilio karna)
- Purple-bearded bee-eater
- Red-casqued hornbill
- Reticulated python
- Rhacophorus tree frog (foam-nest tree frog)
- Rufous fantail (rufous-bellied fantail)
- Spectral tarsier (Sulawesi or Celebes tarsier)
- Sulphur-crested cockatoo
- Tree nymph butterfly (Hestia)
- Wallace's fruit bat (flying fox)

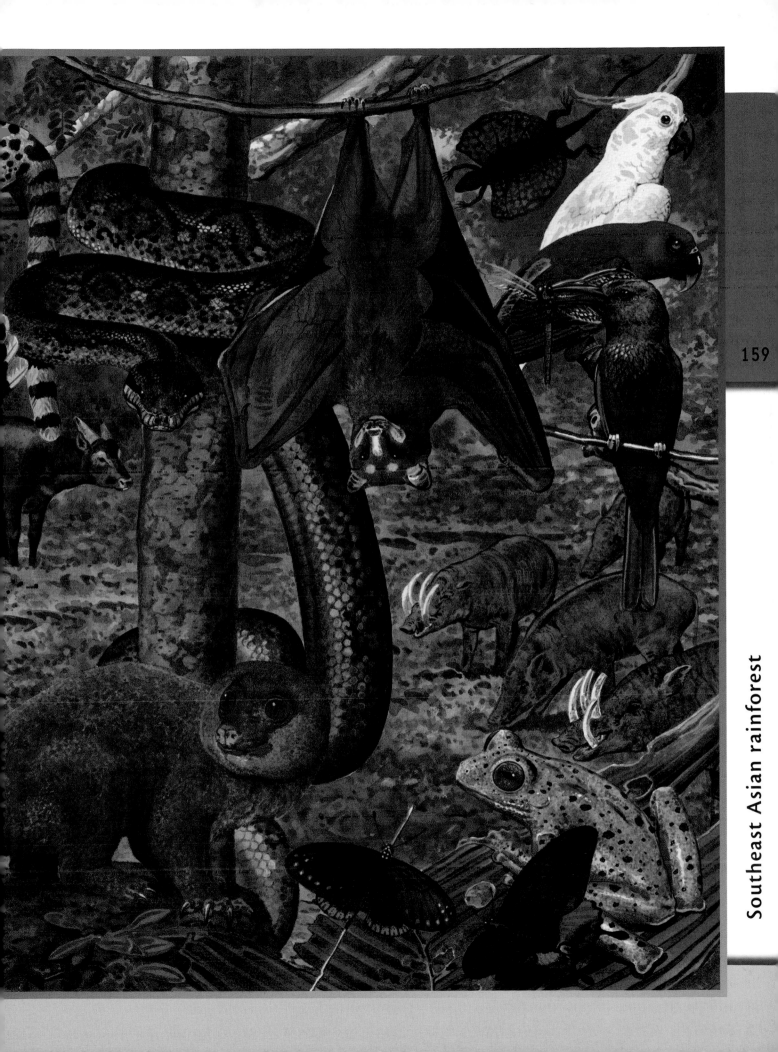

Among the tree trunks

Sulawesi (Celebes) is part of the once-vast rainforest belt across Southeast Asia.

Far above the ground, the sunlit canopy or 'roof' of the rainforest is alive with chattering monkeys, squawking birds, and butterflies flitting among the leaves, blossoms and fruits. The canopy shades out the sun and so the forest floor is usually cast into deep gloom. However some creatures walk among the dead leaves or creep between the tree trunks. The lack of light means little undergrowth and predators such as snakes and the cat-like linsangs have a clear view, especially during daytime. Many creatures, from bats to birds, can fly away from danger. Even a type of lizard, the flying dragon, is able to glide on its wing-like flaps of skin.

Spectral tarsier

Tarsiers are cousins of the bushbabies from Africa. They live a similar lifestyle, bounding between branches at night like spring-loaded toys in their search for small animal prey. They grab and eat beetles, cockroaches, caterpillars, crickets, and other insects, also small lizards, baby birds in their nests and even tree scorpions. The tarsier's head and body measure about 10–12 cm long. Its back legs are half as long again and also very muscular for its great leaps.

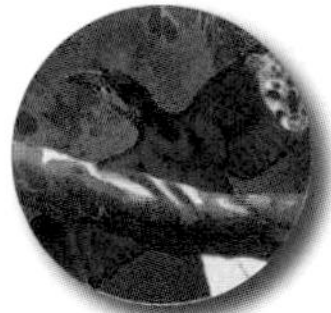

Red-casqued hornbill

The huge beak is made mainly of spongy bone filled with 'bubbles' of air. So it is very light and used delicately like giant tweezers.

Crested mynah

Mynahs are types of starlings. They hop through the forest in flocks, in search of caterpillars, termites, bugs, ants, and similar small insects.

Flying dragon

Not a real dragon, nor a real flier, the flying dragon is actually a gliding lizard. It eats small insects such as ants and its skin 'wings' are held out by extra-long rib bones.

Reticulated python

This is probably the world's longest snake, reaching some 10 m. Its complicated brown-grey pattern gives perfect camouflage against the mottled tree bark.

Tree nymph butterfly

This ghostly pale butterfly with its weak, fluttering flight has become linked to legends of wood nymphs and other spirits of the forest.

Anoa

Anoas are a very rare, shy and elusive type of wild cattle. They are the smallest buffaloes, standing about 90 cm tall at the shoulder. In the past they were hunted for their horns, meat and their very thick skin which makes extremely tough leather. Sadly some illegal hunting still goes on. Anoas are also threatened as their forest home is cut down for timber and farmland.

A tarsier can easily leap 20 times its own body length – equivalent to five times the world long-jump record for a human.

Sulphur-crested cockatoo

With its distinctive long yellow crest, this cockatoo often comes to parks and gardens. It also raids farm crops. A cockatoo usually raises its crest when it is curious, aroused or excited, and lays it flatter behind the head when it is at rest. (See also page 166.)

Wallace's fruit bat

The fruit bats are very different from the smaller, more widespread, insect-chasing bats. Fruit bats have long snouts so their faces resemble dogs or foxes – this is why they are also known as flying foxes. Most types roost by day in large groups in the trees. They hang upside down by their feet and flap their part-folded wings to keep cool. At dusk they take off in a chattering cloud to search for ripe fruits and other soft plant food. There are more than 130 types of fruit bats across Africa, Asia and Australia. Wallace's fruit bat has a head and body about 20 cm long and a wingspan of around 40–45 cm.

Muller's parrot

The parrots are tropical forest birds, clambering through the trees rather than hopping or flying. They use the beak as a third foot to grab branches, and then use the foot as a second beak to hold and manipulate food when eating.

Babirusa

This rare wild pig has very unusual tusks, which are really extra-large teeth. The upper two grow the 'wrong way', upwards and out through the top of the snout. The lower two grow in their usual way, which is also upwards.

Dwarf cuscus

Cousin of the possum, the cuscus is a marsupial (pouched mammal). It lives alone and feeds mainly on fruits, soft leaves and small animals. Its prehensile (grasping) tail curls around branches and has furless skin for a good grip.

Celebes macaque

This macaque monkey has a head and body length of 50 cm and a distinctive head crest of stiff, upright hairs. It is a very general feeder on fruits, shoots, roots and small animals such as insects, mice and birds – in fact, whatever it can lay its hand on.

Rhacophorus tree frog

These tree frogs have huge feet with sucker-tipped toes to grip wet, slippery leaves and bark. They rarely come down to the ground, even to lay their spawn (eggs) in a pool. The frog whips up its body fluid to make a foam nest in the tree for its tadpoles.

TREES AS HIGHWAYS

Tree trunks are highways between the forest canopy above and the bushes, undergrowth and leaf litter perhaps 50 m below. Monkeys, tarsiers and cuscuses have strong hands and feet with grasping fingers and toes to grab smaller branches. Snakes like the python coil around the trunk and slither up or down. Flying lizards and flying squirrels swoop from one tree to land lower on another trunk, then race up the bark with their sharp-clawed toes to set off on another glide.

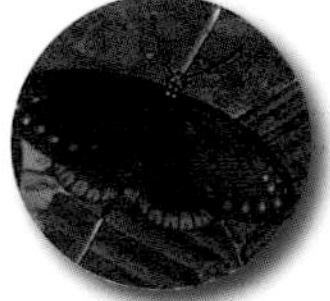

Blue-trimmed crow butterfly

Crow butterflies have a dark, shiny, blue-black colouration similar to their bird namesakes.

Paris peacock butterfly

This beautiful butterfly has a deep, glossy, greenish-blue sheen that glows in the gloom.

The largest fruit bat, and the biggest bat of all, is the greater fruit bat of South and Southeast Asia. Its wings are more than 1.5 m from tip to tip.

Rolling and tumbling

Bee-eaters are slim birds with bright feathers and long, down-curved beaks. Rollers are slightly larger, more stocky and crow-like, but they too have very colourful plumage. Members of both these groups are skilled, agile fliers. Bee-eaters twist and dart at speed as they chase after flying insects, including bees and wasps. The bird rubs or bashes the insect on a branch to remove the sting before eating it. Rollers are named from their spectacular courtship flights when they roll over and over, tumble and somersault in mid air to impress a mate. All the birds in these groups are found mainly in warmer regions, often in dry, sandy grassland or scrub.

Indian roller
It looks dark while stood on its perch, but in flight this roller reveals its dazzling blue wings and tail. It lives in South and East Asia and catches large insects, frogs and lizards.

European roller
A Mediterranean bird, this bulky roller often perches on telephone wires. It swoops down to the ground to catch beetles, crickets and similar large insects, also spiders.

Hoopoe
The hoopoe with its tall head crest cannot be confused with any other bird. It is often heard before it is seen, making the call which led to its name – a soft, three-note 'hoo-poo-poo'.

Red-bearded bee-eater
In the forests of Southeast Asia this large, heavy-bodied bee-eater picks insects from trees, rather than catching them in flight.

Blue-bellied roller
This distinctive roller has a blue-green, deeply forked, swallow-like tail. It inhabits open woodlands in Sudan and Uganda, in Africa.

Lilac-breasted roller
In East and Central Africa, this fine bird is recognized by its lilac chest and long tail streamers. It prefers open bush with isolated trees.

Red-throated bee-eater
Like many bee-eaters, this African type lives mainly in open grassland and savanna, seeking out river banks as nesting sites.

European bee-eater
Common in the Mediterranean area, this bee-eater is also seen in northern France. Favourite foods are bumblebees and dragonflies.

Carmine bee-eater
One of the largest and brightest bee-eaters, this type often rides on the backs of antelopes and cattle. It snaps up the insects they disturb.

To feed itself and its hungry chicks, a bee-eater must catch about 220 bees or similar insects each day. That's one every 4 minutes of daylight time!

Grunts and squeals

- Bush pig (Red River hog)
- Collared peccary (javali, chacharo)
- Giant forest hog (African forest hog)
- Pygmy hog
- Warthog
- Wild boar (Eurasian wild pig)

The farm pig or domestic hog is one of the most numerous and most useful animals in the world. It was probably bred from the wild boar more than 8000 years ago. In addition to the wild boar there are another eight kinds of wild pigs or hogs that dwell mainly in the forests and bushlands of Europe, Africa and Asia. They are ungulates or hoofed mammals related to hippos, camels, goats and cows. Wild pigs are all strong, stocky and heavily built with short legs and bristly coats. Their long, flexible noses or snouts grub in the soil for all kinds of food, and their canine teeth grow extra large to form fierce-looking tusks. *(See also pages 49 and 161.)*

Giant forest hog
This is the largest type of wild pig, weighing over 200 kg and with a head-body length of up to 2 m. It lives mainly in the thick forests and scrubland of Central and East Africa. Giant hogs have broad, flattened snouts and root about for plant food such as leaves, buds, shoots, fruits and berries.

Bush pig
African bush pigs are medium-sized hogs with an upright 'mane' or crest of stiff hairs along the back and tufted ears. They are true omnivores, eating all kinds of foods from leaves, roots and fruits to mice, birds and similar small creatures, and they also scavenge on the bodies of dead animals.

Warthog
Two bony, wart-like lumps on each side of the face give this hog its name. It is found in most of Africa and prefers grassland and bush to forest. Warthogs are preyed on by big cats, hyaenas and jackals but they fight back with great ferocity if cornered. They eat mainly plant food with some grubs and insects.

Pygmy hog
The smallest wild pig is also one of the rarest, officially listed as an endangered species. It is found in the grassy foothills of the Himalaya mountains and measures about 60 cm head-body length with a weight of 7–9 kg. Like most wild pigs it lives in small groups of 10–20, mainly females with their young or piglets.

Wild boar
A very widespread animal, the wild boar lives across Southern Europe, North Africa and Asia, and has been taken to North America. Domestic pigs that live wild tend to 'breed backwards' to look and behave like wild boars. They search in the soil for roots, grubs, seeds, fruits and small animals.

Collared peccary
The three species of peccaries are similar to pigs but live in the Americas. The collared peccary is very adaptable, found in dry scrub, grassland, bush and forest. It also has a wide range from the southern USA to the tip of South America. Peccaries live in small family groups which may form herds of 50 or more.

Farm pigs escape to live wild in some areas. In the Australian outback legends tell of massive boars, big as ponies, with upright crests of hair along the back – 'razorbacks'.

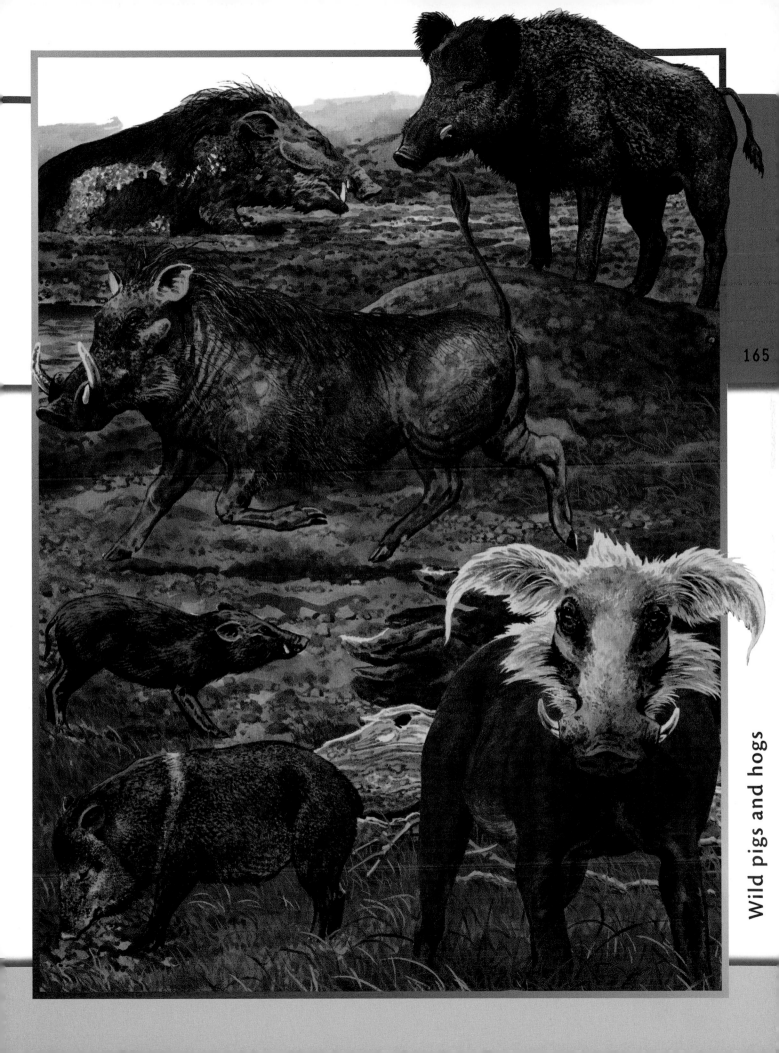

Colourful, curious and clever

- Gang-gang cockatoo
- Hyacinth macaw
- Kea
- Long-billed corella
- Palm cockatoo
- Pesquet's fruit-eating parrot
- Pink cockatoo
- Red-tailed black cockatoo
- Sulphur-crested cockatoo
- Yellow-tailed black cockatoo

Parrots are lively, inquisitive and intelligent – for birds. The group includes cockatoos, parakeets, lovebirds, macaws, conures and lories (lorikeets). A typical parrot has colourful feathers, a large head, big beady eyes and a strong, hooked beak to crush even the hardest seeds. The beak is also used as an extra foot to clamber about in trees. Parrots make many calls and some mimic other sounds, including the human voice. The combination of colour, inquisitive behaviour, copying our speech, the ability to learn tricks, and being long-lived have all made parrots popular pets. But many are still illegally captured and several types are endangered. *(See also page 148.)*

Palm cockatoo
Unusually dark for a cockatoo, this large species has a tall head crest and huge bill to crack open even extremely hard palm nuts and similar seeds. It inhabits tropical forests.

Gang-gang cockatoo
This male of this small grey cockatoo species makes a call like a rusty, creaking gate hinge. In the breeding season he also has a bright red head.

Sulphur-crested cockatoo
This large yellow-crested bird forms flocks to feed on seeds and fruit. It is seen in parks and gardens in north and east Australia and soon becomes tame at the bird table.

Long-billed corella
This cockatoo is unusual because it spends much time on the ground, digging for roots with its long, strong, pale beak.

Hyacinth macaw
The world's largest parrot, from tropical South America, this massive macaw is under threat from unlawful collecting.

Pink cockatoo
The white plumage of this tropical bird is 'shot through with soft pink flush'. Its head crest displays bands of scarlet and yellow when spread.

Kea
The kea is named after its piercing call. It lives in New Zealand and uses its long upper beak to tear flesh from fruits – and to rip flesh off dead animals.

Yellow-tailed black cockatoo
Dark except for its long, yellow-edged tail and yellow cheeks, this parrot from Tasmania and south-east Australia has a weird wailing call.

Red-tailed black cockatoo
These noisy birds gather in flocks of 200 or more. The scarlet patches on the tail shine brightly as the cockatoo flaps along slowly.

Parrots are among the longest-lived of all birds. Larger kinds reach 40–50 years of age. In captivity some have lived for more than 80 years.

Edging toward extinction?

- Black rhino (hook-lipped rhinoceros)
- Indian rhino (greater one-horned rhinoceros)
- Javan rhino (lesser one-horned rhinoceros)
- Sumatran rhino (Asian two-horned rhinoceros)
- White rhino (square-lipped rhinoceros)

Rhinos look like hippos or elephants, but they are close relatives of tapirs and horses. These massive hoofed creatures are the tanks of the animal world, with extremely thick skin, bulky bodies and the long horn (or horns) on the nose. Both male and female have horns. A rhino has small eyes and sees poorly. But its hearing is very good, the large ears swivelling to pick up distant sounds, and its nose is even more sensitive. All five kinds or species of rhinos are threatened, mainly by loss of their natural habitat. But they are still hunted for their horns. The horn may be made into a dagger handle or ground up as a medicine (even though it has no healing effect).

Javan rhino

This is the rarest rhino and one of the world's most threatened animals. It is now found only in wildlife reserves in the west of the island of Java, where its population numbers tens rather than hundreds. Its main habitats are rainforests but these are being felled for timber and to make way for farmland.

Sumatran rhino

The Sumatran rhino of Southeast Asia has two horns, like the two African types. It is the smallest rhino, up to 3 m in head-body length and 750 kg in weight. It also differs from other rhinos because it is partly covered in long hair. It is a rare rainforest animal, with a total number of only 100–200.

Indian rhino

The Indian rhino prefers open scrub and grassland to forests. Its heavy skin, almost 2 cm thick, is divided into distinct plates which give it a suit-of-armour appearance. Also the lumps in the skin make it look as if it has been bolted on! The 1500 or so surviving Indian rhinos live mainly in Bengal, Assam and Nepal.

White rhino

The largest rhino reaches a length of 4.2 m, with a shoulder height of 1.9 m and a weight of 3.5 tonnes. The 'white' does not refer to the colour, which is pale grey. It means 'wide' from this rhino's broad, squared-off snout. White rhinos live in dry bush and grassland across Africa, especially in the south.

Black rhino

Black rhinos, which are actually grey-brown, range widely across Africa. But they have suffered badly from the traps and guns of poachers, being completely wiped out in some areas. The black rhino's long, flexible upper lip can grasp leaves and shoots as it browses on rainforest trees and scrubland bushes. This type of rhino is often active by night, unlike its relatives. Rhinos tend to live alone except for a mother with her baby. (White rhinos sometimes form small groups as they search for food.) One of the main problems facing all rhinos is that, like many large animals, they breed very slowly. The female usually has one young every two to four years. This means rhino numbers take a long time to build up.

Rhino horns are not made of real horn but of very tightly-packed hairs. The largest belonged to a white rhino and was more than 1.5 m long.

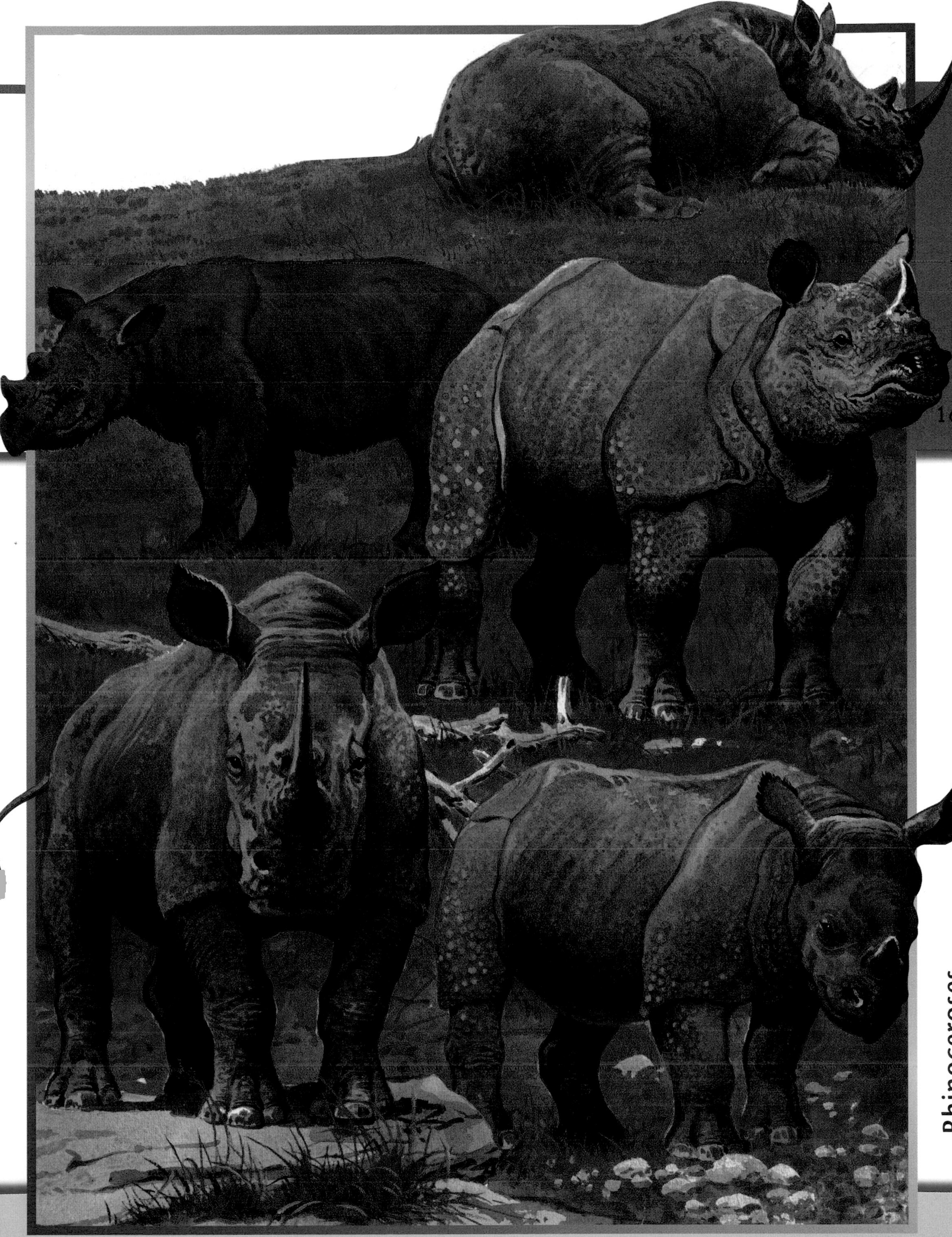

Forests in the clouds

- Black-mantled goshawk
- Black-spangled tree frog
- Cassowary
- Channel-billed cuckoo
- Emerald monitor
- Graceful honeyeater
- Green figbird
- Helmeted friarbird
- Hooded pitta
- Little king bird of paradise
- Long-nosed echidna (spiny anteater)
- Magnificent bird of paradise
- Miliana moth
- New Guinea bandicoot
- Palm cockatoo
- Queen Alexandra's birdwing butterfly
- Raggiana bird of paradise
- Red-bellied pitta
- Red-browed fig parrot (quarrion)
- Red-crowned fruit dove
- Spotted cuscus
- Spot-winged monarch butterfly
- Striped possum
- Tree kangaroo
- Tube-nosed bat
- Ulysses butterfly (Papilio ulysses)
- Victoria crowned pigeon
- White-marbled gecko
- Yellow-bellied sunbird

Feathers fallen from paradise

- Blue (Prince Rudolph's) bird of paradise
- King bird of paradise
- King of Saxony's (Albert's or enamelled) bird of paradise
- Magnificent riflebird
- Princess Stephana's (Stephanie's) bird of paradise
- Raggiana bird of paradise

European naturalists first ventured into the tropical forests of New Guinea in the late 18th century. They were so astonished by the incredible colours and shapes of the feathers that they believed the birds had 'fallen from paradise'. Each newly discovered species was named in honour of a European royal or noble person of the time, such as Princess Stephana, the King of Saxony and Count Raggi. In fact only males (shown here) display such amazing plumes, to impress partners at breeding time. Females are mostly drab brown or green for camouflage, since they must incubate (sit on and keep warm) the eggs and feed the chicks.

Blue bird of paradise To attract his mate the blue bird of paradise hangs upside down from a twig, sprays out his wing and tail feathers and arches his two long tail plumes into an M shape. Then he sways to and fro and makes a courtship call that sounds like the cough and splutter of an old motorcycle engine.

King bird of paradise Like the other 41 species in the bird of paradise group, the king lives mainly in rainforest. It feeds chiefly on fruits, along with some insects, frogs and other small animals. During his mating display the male holds out his wings and vibrates them like a fast-shaking fan.

Princess Stephana's bird of paradise This bird of paradise is found unusually high for the group, up to 3000 m in the remote, cooler, cloud-forest mountains of south-west New Guinea. Most female birds of paradise lay two or three eggs, but the Princess Stephana female usually produces only one.

Raggiana bird of paradise Many birds of paradise are threatened by felling of their rainforest homes for timber (the felling is sometimes done illegally). The Raggiana or Count Raggi's species copes with a wider range of habitats and has spread into scattered lowland woods and even rural gardens.

King of Saxony's bird of paradise Many birds of paradise have long, ornate plumes (slim, ribbon-like feathers). But they are usually on the tail or wings. Only the King of Saxony's bird of paradise has them on the head. It is also one of the smaller types, only 22 cm from bill to tail tip (excluding the plumes).

Magnificent riflebird Riflebirds are members of the bird of paradise group. This type is named after its popping, gun-like call. Like the Victoria riflebird, it lives in the rainforests of north-east Australia as well as New Guinea. Birds of paradise make a raucous squawking similar to their close cousins, the crows.

The magnificent riflebird is one of the few birds of paradise not named after a king, prince or similar titled person.

A bower for courtship

- Archbold's bowerbird
- Green (spotted) catbird
- Golden bowerbird (male and bower in background)
- MacGregor's (crested) gardener
- Tooth-billed bowerbird (tooth-billed catbird)

Bowerbirds are medium-sized, eat mainly fruit, and live in the forests of New Guinea and Australia. The male attracts or courts the female in an extraordinary way by building a bower. This varies according to the kind or species of bowerbird. It may be a simple mat of leaves and moss, or a pile of twigs, or a large and elaborate structure shaped like a tower, tent, maypole or walk-along 'avenue'. The male may even decorate and paint his bower with bright colours. He then dances, shows off his feathers and calls from his bower, to bring a female near. But the bower is not a nest. After mating the female leaves to build her nest in a bush and raise the chicks on her own.

Green catbird
Catbirds are types of bowerbirds but they do not build bowers. They are named from their miaowing, cat-like calls. (They are different from American catbirds, which are types of mockingbirds.) The green catbird of New Guinea and north-east Australia 'mews' at dawn, then forages among trees for fruits.

Golden bowerbird
The golden bowerbird's huge twig-and-stick bower is decorated with colourful mosses, flowers and fruits. It has a maypole-like shape. The finished structure may be as tall as a person! These bowerbirds live in a restricted area in the mountain tablelands of northern Queensland, Australia.

MacGregor's gardener
This bird builds a relatively simple bower. It consists of a vertical stick with twigs, stalks and similar decorations piled up around it. The final effect looks like a person-sized bottle-brush! The MacGregor's gardener is recognized by its thick head crest of golden-orange feathers.

Tooth-billed bowerbird
The dull browns of this Australian species make it hard to see in the forest. But it can be heard – its loud song includes sounds copied from other birds and animals like crickets. Its wavy-edged beak cuts off fresh leaves, both to eat and to decorate a clearing or 'stage' in the forest where the male courts.

Archbold's bowerbird
This bowerbird was discovered quite recently, in about 1940. It lives in mountain forests in New Guinea, at heights of up to 4000 m. Each male builds a mat of grasses and ferns, decorated with snail shells and the shiny wing cases of beetles. He then sings from a perch just above this bower.

LESS MEANS MORE
There are 18 species of bowerbirds. Those birds with the dull feathers build big, complicated bowers. They even decorate them with colourful petals, shells and berries, and smear on 'paint' of natural coloured juices. Males with the brightest plumage make smaller, simpler bowers.

Each type of bowerbird builds its own distinct bower. But young males make small, poorly crafted versions. Their bowers get bigger and better with experience over the years.

Clever cat-like killers

- African civet (African genet)
- African linsang
- Celebes palm civet (giant or brown palm civet)
- Giant genet
- Masked palm civet

Civets and genets are active hunters that look like a combination of cat and weasel. But they form their own mammal group, the viverrids, with about 35 species. Most are found in Africa or Southeast Asia, with the common genet ranging as far north as France. These stealthy predators have pointed faces and long tails, and many have striped or spotted fur. They hunt at night, often leaping through trees to seek out birds, lizards, squirrels, small monkeys and similar prey. Civets make a strong scent called musk. They mark out the boundaries of their territories with it, as a a sort of invisible signpost saying 'Keep out!' to others of their kind.

Giant genet
A secretive creature of the rainforests in Uganda and Zaire, the giant genet has heavily spotted fur. This type of pattern is called disruptive camouflage. It helps to break up the body outline as the genet slinks through the undergrowth, slipping in and out of the shadows and dappled patches of moonlight.

African civet
This is one of the larger viverrids, with a body and tail some 1.3 m long. Found across much of Africa, it is boldly marked and has a crest of raised hairs along the centre of its back and upper tail. Less of a climber than others in the group, it forages on the ground for small animals and occasionally fruit.

African linsang
This slender linsang is about the size of a pet cat. Indeed some linsangs, and especially genets, are kept as semi-tame pets which help to get rid of mice, rats and similar pests. The African linsang spends much time in trees, even building a nest there of twigs and leaves. Birds, lizards and insects make up most of its food.

Celebes palm civet
This very large and rare civet lives only on the island of Sulawesi (Celebes) in Southeast Asia. It is about 1.5 m in total length, including the tail, but is still at home in trees. Like cats, palm civets have retractile claws. This means they can be pulled into the toes to stay sharp.

Masked palm civet
This civet is grey or grey-brown all over except for its face markings. These have a mask-like pattern of black and white, resembling a badger, skunk or raccoon. Masked palm civets are found across a wide area from India east to China and south to Southeast Asia. People have also brought them to Japan. They have a mixed diet including mice, rats, insects and other small animals, and also plant shoots, soft roots and juicy fruits.

The similarity between the masked palm civet and the skunk goes further than face markings. The scent glands around the civet's rear end which produce the musk-based scent are large and their oily product is very strong-smelling. If threatened, the civet turns around and sprays this foul-smelling fluid at its attacker.

The powerful scent made by civets, called musk, was once collected and used as a basis for making fragrant perfumes.

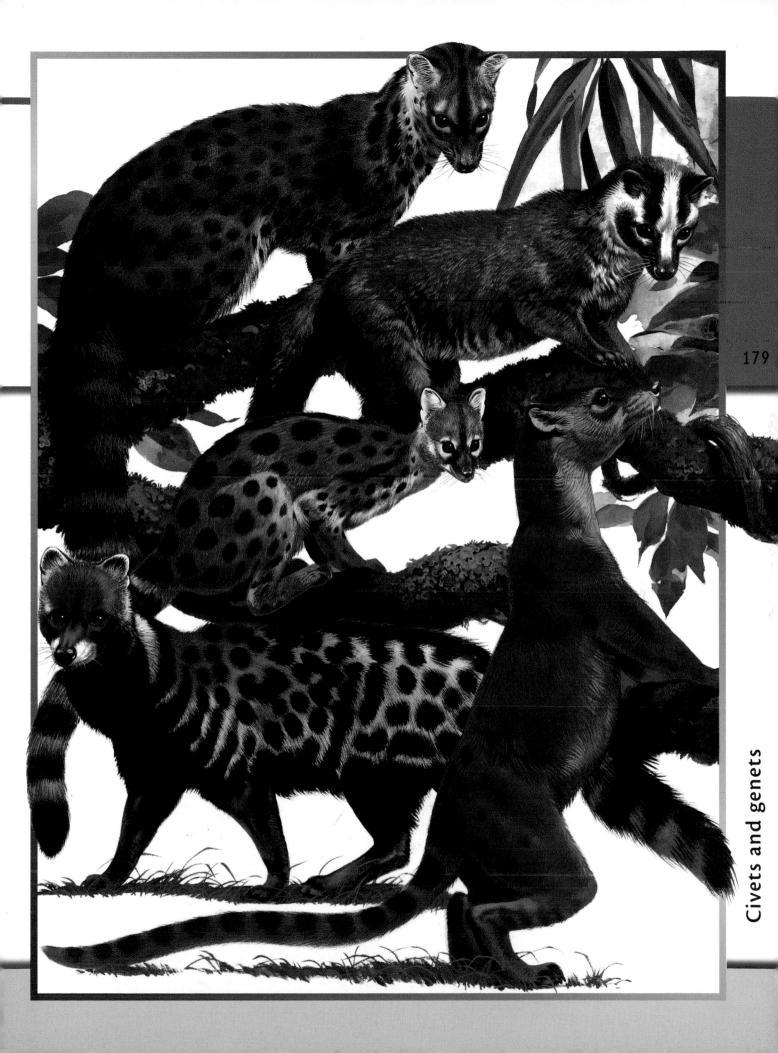

Sturdy birds with crest and tails

- Crested fire-backed pheasant (Bornean fireback pheasant)
- Crested guinea fowl (Zambesi crested guinea fowl)
- Helmeted guinea fowl
- Himalayan monal pheasant (impeyan pheasant)
- Red jungle fowl
- Vulturine guinea fowl

Pheasants and guinea fowl are members of the large group known as gamebirds. Many pheasants have been taken by people ('introduced') to new lands, mainly because of the amazing plumage of the males, or cocks, at breeding time. They have bright flowing tails with brilliant shining colours and patterns. In great contrast the females, or hens, are drab for camouflage. The seven kinds (species) of guinea fowl are sturdy gamebirds from Africa, especially forests and grasslands. Outside the breeding season they often gather in large flocks by day to feed. But when breeding the flocks break up as each pair raises its chicks. *(See also page 156.)*

Crested guinea fowl

This forest guinea fowl is relatively plain, except for its crest which is a mop of narrow curling feathers. Like other guinea fowl and pheasants it feeds by day, mainly in flocks on the ground. Then at dusk it flaps screeching up into a tree, to rest for the night in the branches, safer from predators.

Helmeted guinea fowl

This guinea fowl is named after the horny, hat-like lump on the top of its head. It is a common bird of scrub and grassy plains in many parts of Africa. Like the red jungle fowl it has been domesticated and bred as a source of food. It eats a wide variety of insects, worms, seeds, buds and shoots.

Vulturine guinea fowl

This is the largest of the guinea fowl, at more than 60 cm from beak to tail. It inhabits forests, scrub and bush. Parts of its head and neck are bare, featherless skin which make it resemble a vulture. This is the reason for its name. But lower down it has the usual fine display of shiny feathers.

Himalayan monal pheasant

This male pheasant has the most iridescent or shiny plumage of any bird, in glittering shades of blue, green, yellow, purple and orange. He shows it off by flying and jumping around the female, wings and tail spread. These birds live in open woods in West Asia and northern India.

Red jungle fowl

Originally from Southeast Asia, the red jungle fowl is the wild ancestor of our domestic chickens or hens, in all their different breeds and varieties. It lives in forest and scrub and feeds by pecking and scratching on the ground for a huge range of foods. Shown here is the male (cockerel or rooster).

Crested fire-backed pheasant

Found on the Southeast Asian islands of Borneo and Sumatra, the crested fire-backed pheasant is named from the patch of brilliant red on its back. This contrasts greatly with its otherwise dark blue and brown feathers. It also has an unusual crest with a fluffy tip.

There are more than 200 official breeds or varieties of domestic chickens. All are descended from the red jungle fowl.

Our closest cousins

- Chimpanzee (common chimpanzee)
- Gorilla
- Lar gibbon (common or white-handed gibbon)
- Orang (orang-utan)
- Siamang (black gibbon)

Apes are cousins of monkeys – and ourselves. An ape has a large head and brain relative to its body size, forward-facing eyes, front limbs longer than back limbs, and no tail. There are two main ape groups. The great apes include just four species – the orang of Southeast Asia, and the gorilla, common chimp and pygmy chimp (bonobo), all of Africa. Their hands have very moveable thumbs that can grip and manipulate. The lesser apes are called gibbons. There are nine species, all found in the forests of Southeast Asia. Gibbons are incredibly agile climbers. They swing through the branches using their long, muscular arms.

Orang
Orangs live only in the thick forests of Sumatra and Borneo. They are rare, shy and seldom seen. Their bright reddish-brown fur grows long and shaggy with age. Orangs spend many hours feasting on soft fruits such as figs and mangos. They also eat leaves, shoots, nuts, soft bark and occasionally small animals.

Lar gibbon
The fur of the lar gibbon varies across its wide range of Thailand, Malaysia and Sumatra, from black in some areas to buff, brown or red in other regions. But its hands and face-ring are always white. Gibbons eat mainly ripe fruits, also soft leaves and soft-bodied animals such as insect grubs.

Gorilla
The huge and muscular gorilla is a peaceful animal which eats almost entirely plant food. Gorillas live in small family groups in the dense African forest, feeding mainly on leaves and stems. They spend most of their time on the ground. At night they generally sleep in tree nests made by bending branches together.

Siamang
With a head-body length of almost 100 cm, this is the largest gibbon. It inhabits the forests of Malaysia and Sumatra. The male and female make loud calls and often sing together in a 'duet'. Unusually for a mammal, when the baby siamang is about one year old, the father takes over from the mother and looks after it.

Chimpanzee
Although called the 'common' chimpanzee, this ape is becoming scarce in its natural habitat – the savannahs, bush and forests of Central Africa. Chimps face destruction of their home areas and illegal trapping for the trade in exotic pets. They are mainly plant-eaters but also take some animals, especially termites, ants and certain kinds of caterpillars. Occasionally young adult male chimps from a troop form a 'hunting party'. They chase after and kill a larger prey such as a monkey, bird, small deer or even a chimp from a neighbouring troop. A baby chimp can do little except cling to its mother and feed on her milk. As it grows it begins to play with other youngsters in the troop. But it stays with its mother for 6–7 years.

A large male gorilla grows to 1.9 m tall, weighs up to 200 kg, and is as strong as three adult humans.

Too big to fly

- Bennet's (dwarf) cassowary
- Cassowary
- Emu
- Kiwi
- Ostrich
- Rhea

The world's largest and most unusual birds live on the open grassland and scrub of the southern continents. They cannot fly – their wings are too small and their bodies are too big and heavy. They are the ostrich of Africa, the emu of Australia and the rhea of South America. They are tall and strong with powerful back legs, and they can run rapidly from danger or kick and slash enemies. The cassowary and kiwi skulk in dark, dense forests. All of these flightless birds lay large eggs. In the ostrich, emu and rhea several females put their eggs in one nest and the male guards them. The chicks huddle together into a small flock as soon they hatch.

Rhea
Smaller than the ostrich and emu, the rhea is still the largest bird in the Americas. It roams the pampas (grassland), often gathering in large flocks of a hundred or more birds. At breeding time the male rhea gathers a group or harem of up to a dozen females, then digs a hole for them to lay their eggs.

Emu
The emu is almost as large as the ostrich, at nearly 2 m tall. These big, tough birds live in many parts of Australia, where they feed mainly on grasses, fruits, flowers and seeds. Emus are usually on the move. They search the dry outback for an area where it has recently rained and plants are growing.

Bennet's cassowary
This type of cassowary is about 100 cm tall. It has a fearsome reputation as a very dangerous bird. It may attack anything or anyone who comes too close to its nest, kicking out with the sharp claw on each foot. These birds live in scrub and bushy country in the mountains of New Guinea.

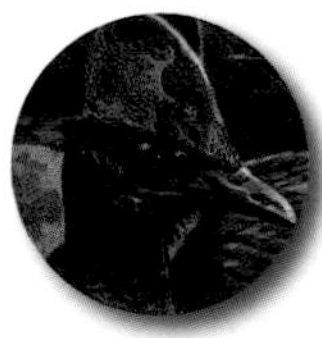

Cassowary
Found in parts of northern Australia and New Guinea, the cassowary has colourful, turkey-like wattles (flaps of skin) hanging from its head and neck. To attract a female at breeding time the male shows off these wattles. He also balloons out his throat pouch to make strange, deep, booming sounds.

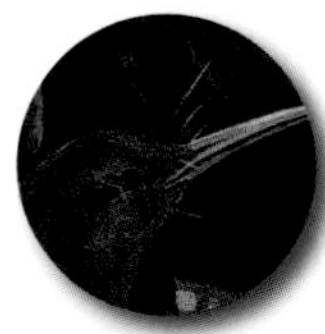

Kiwi
Symbol of New Zealand, the kiwi is well known yet seldom seen. Its feathers are so thin and fine that the bird looks like a ball of fur. Kiwis creep about in the forest at night, pecking and poking in the soil for worms and grubs. Compared to its body size the kiwi's egg is the largest of any bird.

Ostrich
The ostrich is the tallest (2.5 m), heaviest (150 kg) and fastest-running (40 km/h) of all birds. Ostriches favour the dry savanna (grassland) of Africa and survive well on a wide variety of foods, from seeds to lizards and frogs. The male shades the eggs and chicks from the scorching sun with his plume-like wing feathers.

The ostrich lays the largest eggs of any bird. Yet compared to the size of the mother bird they are the smallest eggs of any bird. Biggest are the kiwi's (see above).

At the billabong

- Australian pelican
- Black swan
- Budgerigar
- Common brushtail possum
- Copperhead snake
- Eastern rosella
- Emu
- Galah cockatoo (rose-breasted cockatoo)
- Koala
- Kookaburra
- Magpie goose (pied goose)
- Nankeen night heron
- Prettyface wallaby (whiptail wallaby)
- Red kangaroo
- Royal spoonbill
- Southern hairy-nosed wombat
- Spotted-tailed quoll (tiger or native cat)
- Sulphur-crested cockatoo

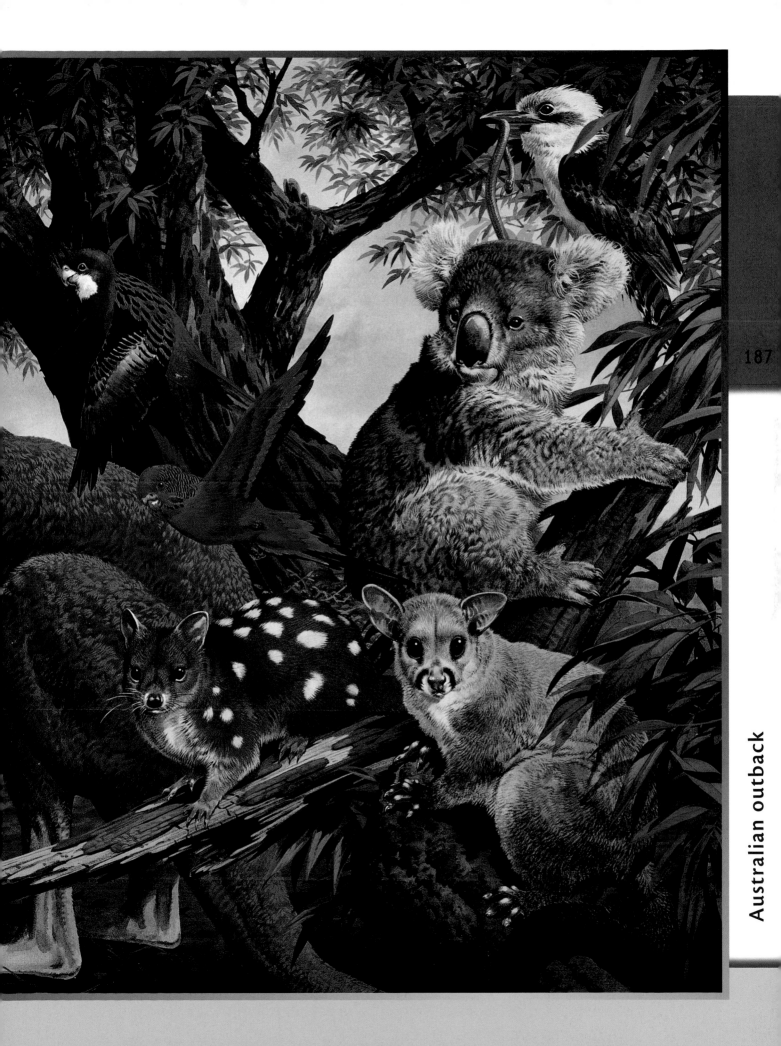

Babies in pouches

- Brush-tailed phascogale (common wambenger)
- Hairy-nosed wombat
- Numbat (banded anteater)
- Quokka
- Red kangaroo
- Red-legged pademelon
- Tasmanian devil
- Yellow-footed rock wallaby

There are about 4080 kinds or species of mammals. Some 270 are marsupials – pouched mammals, named after the marsupium, a pocket or pouch on the front of the female's body. Marsupial babies are born at a very early stage of development, tiny and furless, eyes closed and limbs hardly developed. They wriggle to the mother's pouch where they stay, feed on her milk, grow and develop. Gradually the young marsupials begin to leave the pouch for short periods, although they still return for milk and safety. Marsupials live mainly in the Australian region, with some in South America and the Virginia opossum ranging into North America.

Yellow-footed rock wallaby
This tough wallaby lives in dry, rocky scrub in Central Australia. It eats almost any type of plant and leaps nimbly among the crags.

Red kangaroo
The largest of all marsupials, at 1.8 m tall and 80 kg in weight, only the male or 'boomer' is rusty-red. The female is smaller and known as the 'blue flier' due to her grey-blue fur.

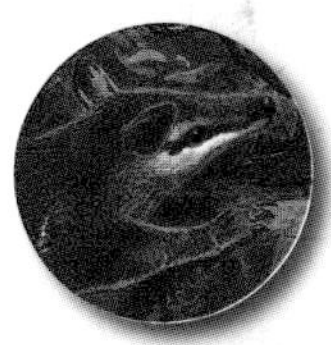

Numbat
In south-west Australia the numbat is like a marsupial anteater. It has a sticky 10-cm tongue and licks up ants, termites and similar tiny food.

Tasmanian devil
The 'devil' is a small marsupial version of the hyaena. Active at night, it hunts small animals and scavenges on the carcasses of larger ones.

Quokka
One of the smaller wallabies, the quokka is 50 cm long with a 30-cm tail. It is found in south-west Australia and eats plants. It has also taken to scavenging from rubbish tips.

Hairy-nosed wombat
Wombats are like marsupial badgers. They live in large tunnel networks and emerge to eat shoots, roots, fruits and other plant food.

Red-legged pademelon
Pademelons are types of heavily built kangaroos which live in forests. The red-legged pademelon is small, with a head and body only 60 cm long. It lives mainly in damp woodlands along the east coast of Australia, foraging for leaves, shoots and similar soft plant food.

Brush-tailed phascogale
There are marsupial versions of many types of mammals. The phascogales are like marsupial squirrels. They are widespread across Australia and have strong gnawing front teeth. Unlike real squirrels they eat small animals such as mice, birds and their eggs, insects and also honey.

A hungry numbat can eat about 20,000 ants in one day.
A red kangaroo can bound along at 50 km/h.

Long-eared, long-legged leapers

- Black-tailed jackrabbit
- Brown hare (common hare)
- Mountain hare (varying or tundra hare)
- Northern pika
- Rabbit (common or European rabbit, cony)
- Red pika
- Snowshoe hare
- Steppe pika

Rabbits and hares live in open, grassy country. With their huge ears, large eyes and twitching noses, they are always alert to danger and ready to escape into their burrows, hide in the undergrowth or race off at great speed. The rabbit itself has been taken to many parts of the world where it thrives – especially Australia. But introduced rabbits not only ruin farmland, they damage the balance of nature in wilderness areas. There are about 45 species in the group. Those with longer legs and longer ears are usually called hares (jackrabbits in North America). Pikas are small vole-like cousins of rabbits. They live mainly in Asia.

Mountain hare
This dappled hare lives mainly in the Arctic and mountain habitats. The more snow that falls in its region, the whiter its coat turns in winter.

Brown hare
Found in many parts of Europe and Asia, the brown hare can cover 3 m in one leap. However it only runs as a last resort, preferring to crouch still until danger has passed.

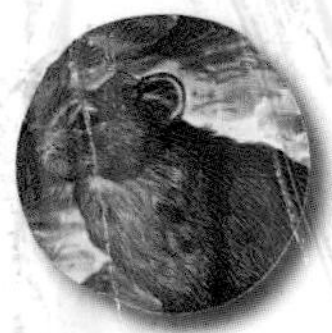

Red pika
Most pikas prefer rocky upland slopes. The red pika lives in the Tien Shan mountain range in central Asia, at heights of up to 4000 m.

Snowshoe hare
The hairy feet of this hare allow it to run over snow without sinking. It lives in the northern forests of North America and is a favourite prey of lynx.

Black-tailed jackrabbit
This North American hare's huge ears detect very faint noises. They also give off heat and help the jackrabbit stay cool in the hot summer sun.

Steppe pika
The steppe pika digs an elaborate network of burrows in the steppes or grasslands of Asia. It retreats underground to escape its many predators.

Northern pika
Pikas are small creatures resembling voles or guinea pigs, with rounded ears and almost no tail. Like rabbits they feed mainly on grasses and other plants. Pikas 'hay-make', gathering grass stems into their burrows during the summer and autumn. They eat this food during the long winter.

Rabbit
The common rabbit is all too common in many places. It came originally from Spain and North Africa but it has invaded many other countries, eating local plants so that other animals starve, as well as destroying crops and digging its home burrows, called warrens. The many types of pet rabbits are bred from it.

The brown hare can race along at speeds of up to 65 km/h.
Large-eared pikas live higher than almost any other mammal, 6000 m up in the Himalayas.

Wide-mouthed night hunters

- Barred owlet-nightjar (Bennett's owl-nightjar)
- Common potoo
- Guachero (oilbird)
- Pennant-winged nightjar
- Standard-winged nightjar
- Tawny frogmouth
- Whip-poor-will

Nightjars are big-eyed, fluffy-looking, graceful fliers with some of the best camouflage in the bird world. Their mottled grey-brown plumage makes them almost impossible to spot as they sit perfectly still by day, among leaves and twigs on the ground or in a tree. At night they glide through the dark, snapping up moths and similar insects in their gaping mouths. Frogmouths are close cousins of nightjars, and as their name suggests, they too have wide, gaping mouths. They also fly at night but are less agile than nightjars, diving down from a perch to catch prey on the ground. Many of these birds have strange calls that sound like engines or machines.

Guachero

The guachero of Central and South America is a close relative of nightjars. It is unusual in many ways. It spends all day deep in a cave, then comes out at night to feed on very oily fruits. The guachero finds its way in total darkness by making loud clicks and hearing the bounced-back echoes – just like a bat.

Common potoo

Potoos are named after their calls and are in the same bird group as nightjars. The common potoo of Central America is another bird with incredible camouflage. When alarmed it stretches upwards on its perch and goes very stiff, so that it looks just like an extra piece of broken branch.

Tawny frogmouth

The tawny frogmouth's plumage blends perfectly with its background as it sits on a tree, looking exactly like an old bark-covered stump. This frogmouth is found in Australia and is large enough to swoop down onto mice and lizards, which it tears up with its strong, hooked beak.

Standard-winged nightjar

The 'standards' of this bird's name are two enormously long wing feathers which may stretch for 50 cm. The courtship flight of this nightjar, which lives in Africa, is very spectacular. As the male flies slowly in circles, his long feathers trail behind like a miniature airplane towing a display banner.

Pennant-winged nightjar

A bird of Central and southern Africa, this nightjar has a long, trailing plume on each wing. The plumes may be twice as long as the bird itself. Only the end of the plume is feathered. Most of it is the wire-like shaft. As in similar nightjars, the male uses them to impress the female.

Whip-poor-will

An American type of nightjar, the whip-poor-will is named after its call. This can be repeated so often for so long that it becomes very irritating to human listeners! The whip-poor-will is fairly common in conifer and mixed broadleaved woods. It hunts close to the ground, snapping up large insects.

The nightjar's song does not sound like a bird at all, but like the chugging of a distant motorbike.

Growls and howls in the wilderness

- African wild dog (Cape hunting dog)
- Bat-eared fox
- Black-backed jackal (silverback jackal)
- Coyote
- Dingo (Australian wild dog)
- Fennec fox
- Grey wolf (timber or white wolf)
- Maned wolf (maned fox)
- Red fox

The dog family includes wolves, foxes and jackals as well as the domestic dog with its hundreds of breeds or varieties. Most foxes live alone but other dogs dwell in groups or packs, usually led by a chief male. They keep in touch and show their mood by barks, howls, whines, growls and many other noises. Cats may be stealthy sprinters, but dogs can run and trot all day on their long, powerful legs and eventually wear down their prey. All members of the group eat meat but many also take a wide range of other foods including grubs, worms, insects and berries. Domestic dogs were probably bred from the grey wolf thousands of years ago.

Grey wolf
Largest of the dog family, these wolves live in forest, scrub and mountains around the northern half of the world. They eat many foods, from deer to rabbits, mice, berries and fruits.

Black-backed jackal
These hardy jackals dwell in dry parts of east and southern Africa. They scavenge on zebras, antelopes and other large prey of big cats, and also kill smaller victims.

Fennec fox
This is the smallest fox. It lives in the Sahara Desert region, where it feeds mainly on ants, termites and other tiny prey. Its big ears help give off excess body warmth in the desert heat.

Maned wolf
The maned wolf's stilt-like legs give it a good view over the tall grasslands of South America. It searches at night for small animals and fruits.

Coyote
Coyotes are still common in some regions of North America, despite years of hunting. They can interbreed with grey wolves and also domestic dogs.

African wild dog
This fierce, large-fanged predator eats almost wholly meat. A pack of wild dogs can easily overpower a wildebeest or zebra.

Dingo
Dingoes may once have been part-domesticated. They now roam wild in the Australian bush. Unusually for dogs, they make hardly any sounds.

Bat-eared fox
This small African fox's huge ears locate its prey of large insects such as beetles and crickets. It can even hear dung-beetle maggots munching!

Red fox
Across Europe and North America the adaptable, clever red fox has taken to living in towns and even cities. It often raids dustbins, even in daylight.

One of the rarest members of the dog family is the African wild dog. There are only about 5000 left, mainly in one wildlife reserve in Tanzania.

High and dry in heat and cold

- Alpaca
- Bactrian camel (Asian or two-humped camel)
- Dromedary (Arabian or one-humped camel)
- Guanaco
- Llama
- Vicuna

Camels are specialized for harsh habitats. The dromedary endures heat and drought in the deserts of North Africa, the Middle East and Australia. The Bactrian camel also survives drought – and the biting winter winds of the Gobi Desert. Most camels today are domesticated by people as beasts of burden and to provide milk, meat, wool and skins. However a few Bactrian herds still roam the Gobi, and dromedaries taken to Australia by people now run wild in the outback. Llamas and alpacas are South American versions of camels, living mainly along the Andes mountains. They are also domesticated. Their wild cousins are guanacos and vicunas.

Alpaca

Like woollier versions of the llama, some alpacas live wild at heights of up to 4800 m in the Andes. Also like llamas, alpacas have been domesticated, although mainly for their fine wool rather than as pack animals. They grow to about 100 cm tall at the shoulder and weigh 60 kg.

Vicuna

The vicuna is the smallest member of the camel group and lives at heights of up to 5500 m, where the air is very thin. Its numbers fell drastically through hunting for meat and its exceptionally fine woolly coat. Since protection in 1969 its numbers have risen to more than 100,000.

Guanaco

Once found in many parts of South America, the guanaco is now limited to Argentina. Like its relatives, it is highly prized for its fur. Guanacos are found in dry grasslands, shrublands and even in forests. They live in small herds of one male with several females and their young.

Bactrian camel

The Bactrian's thick fur helps to keep out the winter cold of the Mongolian high grasslands. The hump of a camel does not store water. It contains fat, which the camel can use as a food store when plants are scarce. However the fat can be broken down by the camel's special body chemistry to produce water too.

Llama

The llama is found in many regions of mountain grassland and scrub along the Andes. Like the camels it has a long history as a domesticated animal – for transport, as well as for its fur, milk and meat. Llamas have been taken to wildlife parks in many other continents and are also bred as large, pony-like pets.

Dromedary

There are now two breeds of this camel species. One is tall and strong, and still used as a pack animal in remote places where there are few roads. It seems to stride along slowly, but it can walk all day and cover 30 km even with a heavy load. The other breed is smaller and much slimmer and used for racing.

A camel can go for a week without water, then drink 50 litres in just a few minutes.

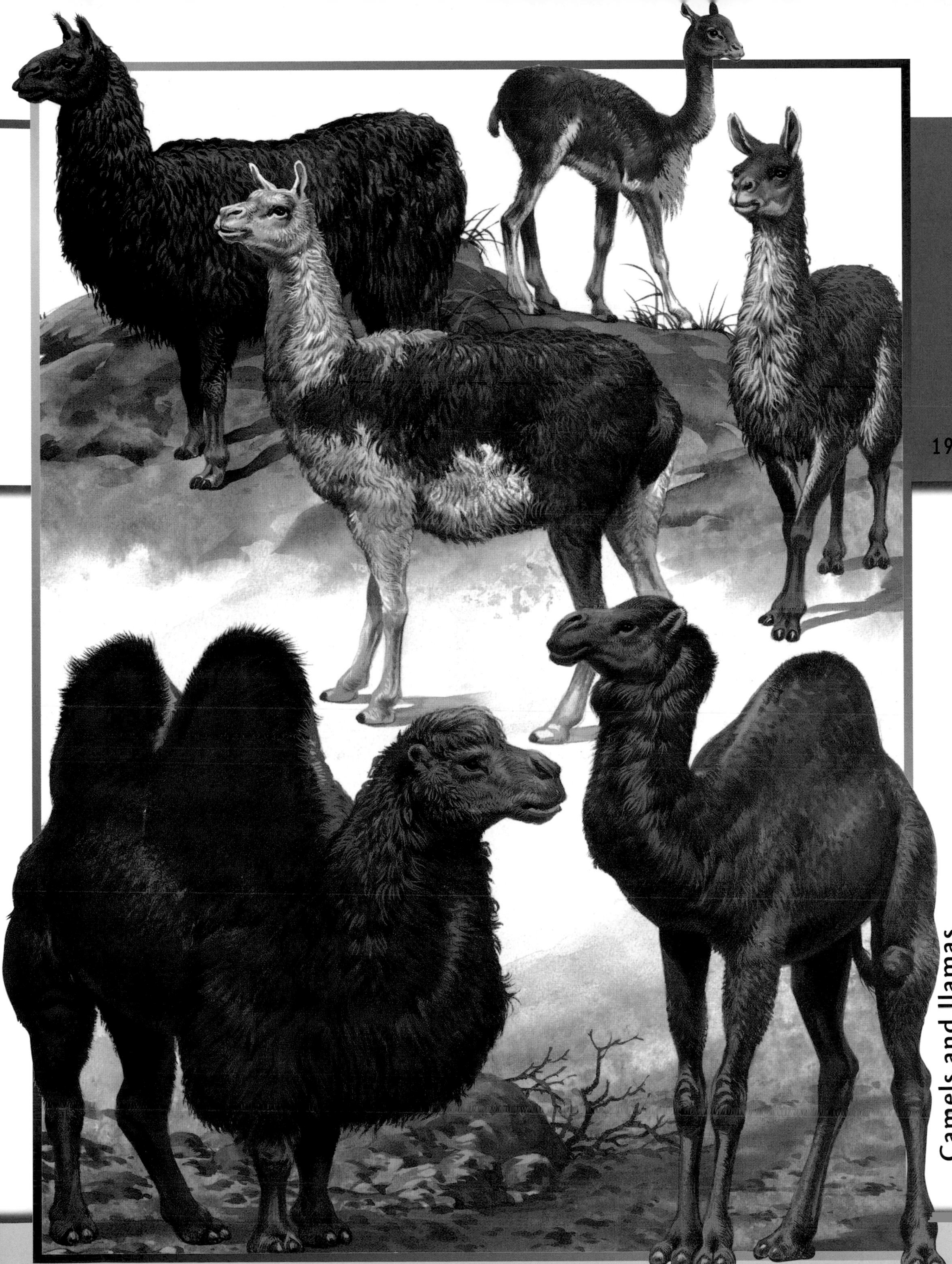

The icy waters

- Adelie penguin
- Chinstrap penguin
- Common seal (harbour seal)
- Crabeater seal
- Emperor penguin
- Gentoo penguin
- Harp seal
- Ice fish
- Leopard seal
- Polar bear
- Ribbon seal
- Southern elephant seal
- Squid
- Walrus

Master hunters of the ocean

- False killer whale
- Killer whale (orca)
- Pilot whale (long-finned pilot whale or pothead whale)

The killer whale is one of the world's supreme hunters. At 9 m in length and 3 tonnes in weight, it is far bigger than any land predator. It's also larger than the biggest flesh-eating shark, the great white. Killer whales are very intelligent. They live in family groups, or pods, and work together to surround prey such as shoals of fish. False killer whales are slightly smaller and do not have such a tall back fin or white patches on the sides of the body. But they too are fearsome hunters of many kinds of prey. Pilot whales grow to about 6 m long. Like the killer whales, they are not really true whales – they are very big members of the dolphin family.

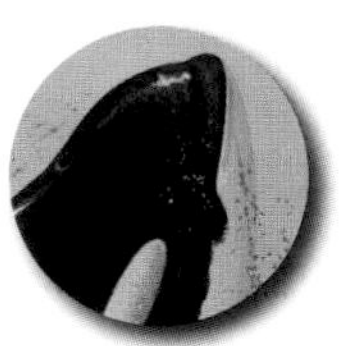

Killer whale

The female killer whale is slightly smaller than the male, at about 6–7 m long. Also the fin on her back is shorter and more curved or crescent-shaped compared to the male's. His tall, pointed fin may be 2 m high – the largest back fin of any dolphin or whale. Killer whales live in all seas and oceans, even in cold Arctic and Antarctic regions. They feed on many kinds of fish, squid and similar prey. They are also the only type of dolphin or whale that regularly hunts warm-blooded victims including other dolphins, also porpoises, great whales, seals, sealions and seabirds such as penguins. Full-grown killer whales have no natural enemies and even in the wild they may survive to an age of 60 years or more. In captivity they are found to be intelligent.

False killer whale

Like the killer whale, the false killer makes a huge variety of sounds. Some are used to communicate with other members of its group. Other sounds work like sonar (sound radar) and bounce off objects. The false killer hears the echoes and so finds its way in dark or muddy water.

TOGETHER IN A POD

A typical killer whale pod has about 10 members. There is usually one large male, three or four adult females, and several youngsters who are both males and females. They may stay together for years. The babies tend to stay and grow up in the pod, generation after generation.

Long-finned pilot whale

Pilot whales are named because they often swim alongside ships and boats. They seem to guide the boat across the sea, in the way that the expert human sailor called a pilot guides ships into a port or through dangerous waters. Pilot whales tend to stay near the coast rather than head for open seas. They live in the North Atlantic and also in all southern oceans. Their main foods are fish and squid. These dolphins stay together in their pod for many years, in a group of six to ten. Sometimes several pods join together for a time to form a larger school. Like all whales and dolphins, pilot whales are mammals and breathe air. So they have to make frequent visits to the surface.

The killer whale is the fastest swimming mammal, powering along at speeds of 55 km/h.

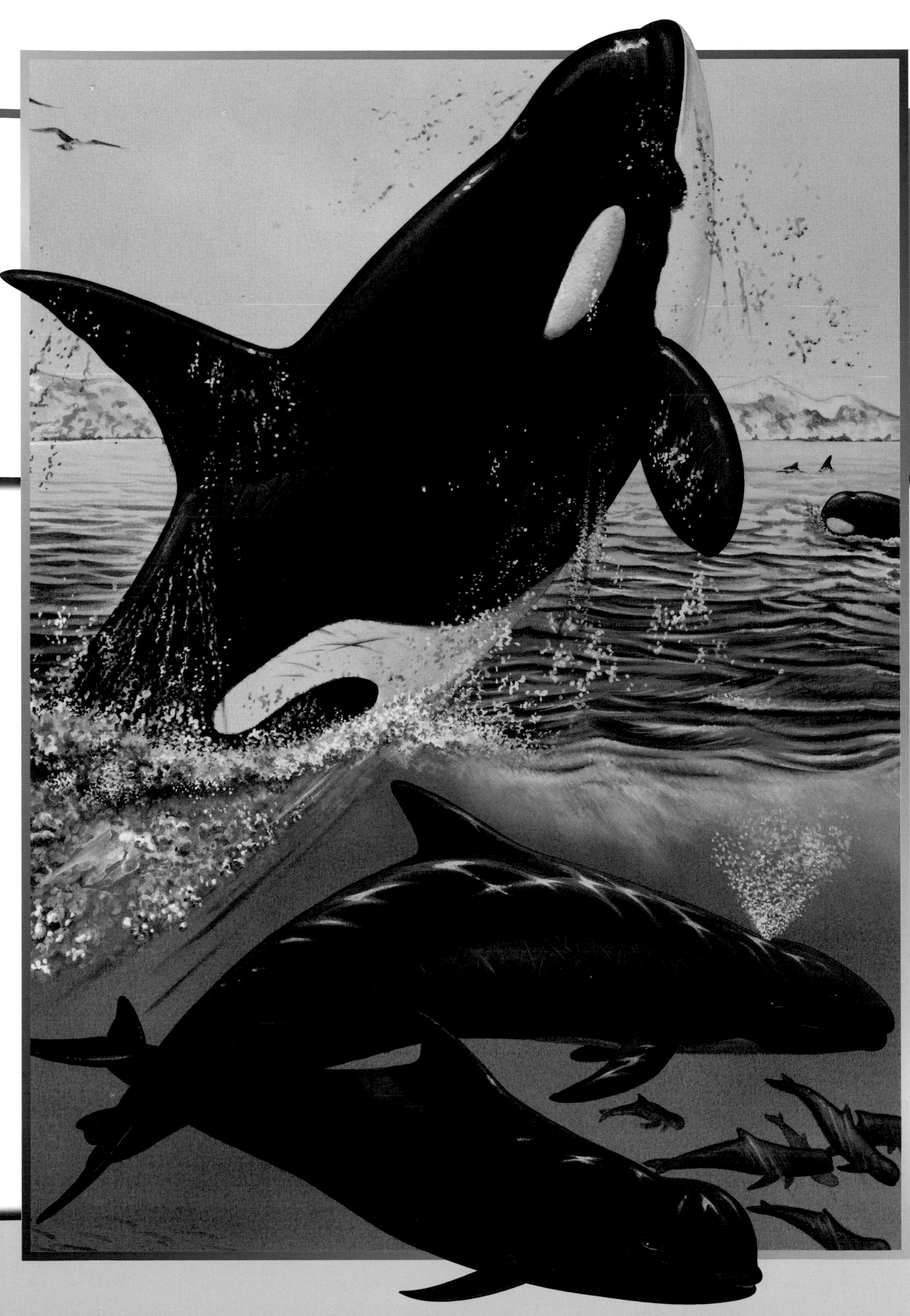

Superb swimmers of colder seas

- Caspian seal
- Common seal (harbour seal)
- Harp seal
- Ribbon seal
- Ringed seal

Seals are superbly equipped for life in water. Most live in oceans but a few, like the Caspian seal, dwell in large lakes and rivers. They swim by bending the body from side to side with a snake-like wriggling motion, the rear flippers working like a fish's tail to thrust the seal forwards. The front flippers trail by the sides of the body or are used for steering and braking. Seals are excellent divers – some types can stay under for more than 20 minutes. But on land they are clumsy and slow, humping along awkwardly (see page 16). The 19 species of seals are found mainly in colder northern and southern waters and feed on fish, squid, shrimps and even seabirds.

Harp seal
A seal of the far north, in the North Atlantic and Arctic Ocean, harp seals rarely come to land. They spend most of their lives swimming or resting on icebergs. Their pups are even born on floating pack-ice. The young pups have soft, silky, pure white fur. The darker markings develop over the first few months.

Common seal
The common seal is one of the most familiar of all seals. It lives in all northern oceans, wandering as far south as the shores of France and Cape Cod in the Atlantic, and California and Japan in the Pacific. As the other name of harbour seal suggests, these seals tend to stay near land and are sometimes seen in ports and harbours and along holiday coasts. They can be inquisitive, coming to ships and quays to look for food. They also swim up rivers into lakes. The common seal is about 1.5–1.8 m in total length and weighs 90–120 kg, although as in other seals, males are slightly larger than females. They feed on fish, squid and shellfish and like other seals they are fairly long-lived, sometimes over 30 years.

Caspian seal
This small seal, about 1.2 m long, is found only in the huge inland saltwater lake called the Caspian Sea in Asia. It breeds on the pack-ice which forms there during the winter. When the water warms up in the spring the seals spread out. They spend most of their time fishing in deeper, cooler waters in the south of the lake.

Ribbon seal
These seals live mainly in the Bering Sea of the North Pacific, where they breed in early spring on the pack-ice. As summer arrives and the ice melts they roam in the open ocean far from land. They live in deeper water than other seals and dive far below the surface to hunt for fish, shrimps and squid.

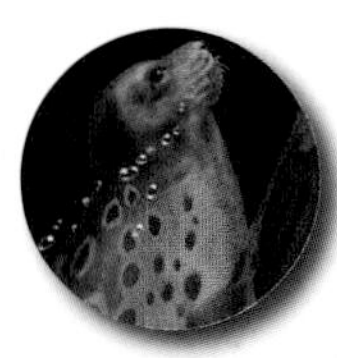

Ringed seal
The ringed seal lives more on its own compared to other seals. It is found in northern seas and also freshwater lakes of Finland and Russia. It makes breathing holes in the ice as this forms in autumn, and keeps them open all winter. Seals have a thick layer of fatty blubber under the skin to keep them warm.

The ringed seal is one of the most common seals, with a world population of more than 5 million – far greater than the number of so-called 'common' seals!

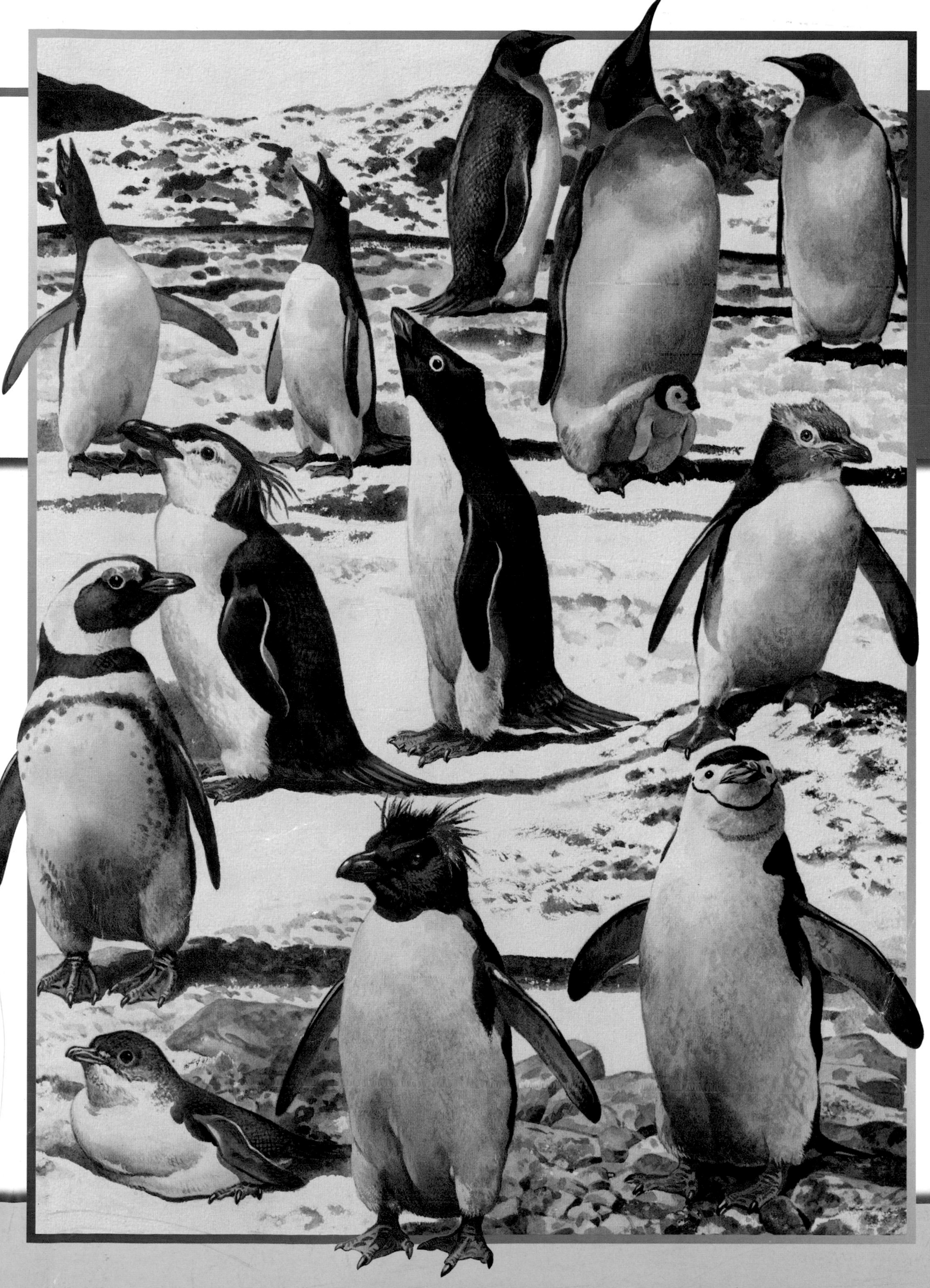

Fish in their millions

- Alewife
- Atlantic herring
- Atlantic menhaden (mossbunker)
- Gizzard shad (Atlantic gizzard shad)
- Pacific sardine

Few fish have affected world history as much as the herring. This smallish, silvery ocean swimmer has been caught in such huge numbers that it has made cities and countries rich. People have even gone to war over who should fish the vast shoals that swim in the North Atlantic. Close relatives of the herring are sardines (pilchards) and anchovies. These are also food fish of global importance. There are about 290 different kinds (species) in the herring group and most live in massive shoals in the open ocean. They feed by filtering tiny animals and plants of the plankton, using long, comb-like parts called rakers on their gills to sieve the water.

Pacific sardine
Each large area of ocean around the world has its own type of sardine. Most are about 20–30 cm long. Like herring, they filter tiny planktonic food and live in gigantic schools which are caught by giant fishing boats. They move with the seasons, away from the tropics for summer and back to warmer waters for winter.

Alewife
The alewife, a type of herring cousin called a shad, migrates like the salmon. It grows up along the Atlantic coasts of North America, then swims up rivers to spawn (lay eggs). It reaches about 40 cm in length. However some alewives stay all their lives in fresh water, mainly in the Great Lakes, and are smaller.

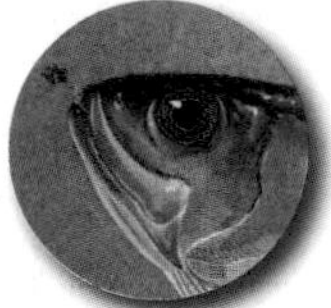

Atlantic menhaden
The menhaden, about 45 cm in length, is another herring relation that swims in great shoals. Huge quantities are caught but they are not usually eaten by people since their flesh is very oily. Instead these fish are used to make fertilizers, fish oils and nutrients added to farm animal feeds.

Gizzard shad
This fish is named after its large gizzard – a muscular bag like the stomach which grinds up its food. Gizzard shads are 50 cm in length and have a very long rear spine or ray on the back (dorsal) fin. Unlike other shads, they stay in the sea to breed and do not swim up rivers.

Atlantic herring
The herring grows to 40 cm long and is pale silver. It is found in shoals several kilometres long with hundreds of thousands of fish. At least it was. So many have been caught by fishing boats that they are now less numerous. Herring are a main link in ocean food chains. They eat tiny plants and animals in the plankton, and they are food for larger fish like marlin and tuna, as well as seabirds, seals, dolphins and many other ocean hunters.

One of the largest herrings is the wolf herring. It grows to 3.5 m in length and has long, sharp teeth as big as a real wolf's fangs – a truly fearsome fish.

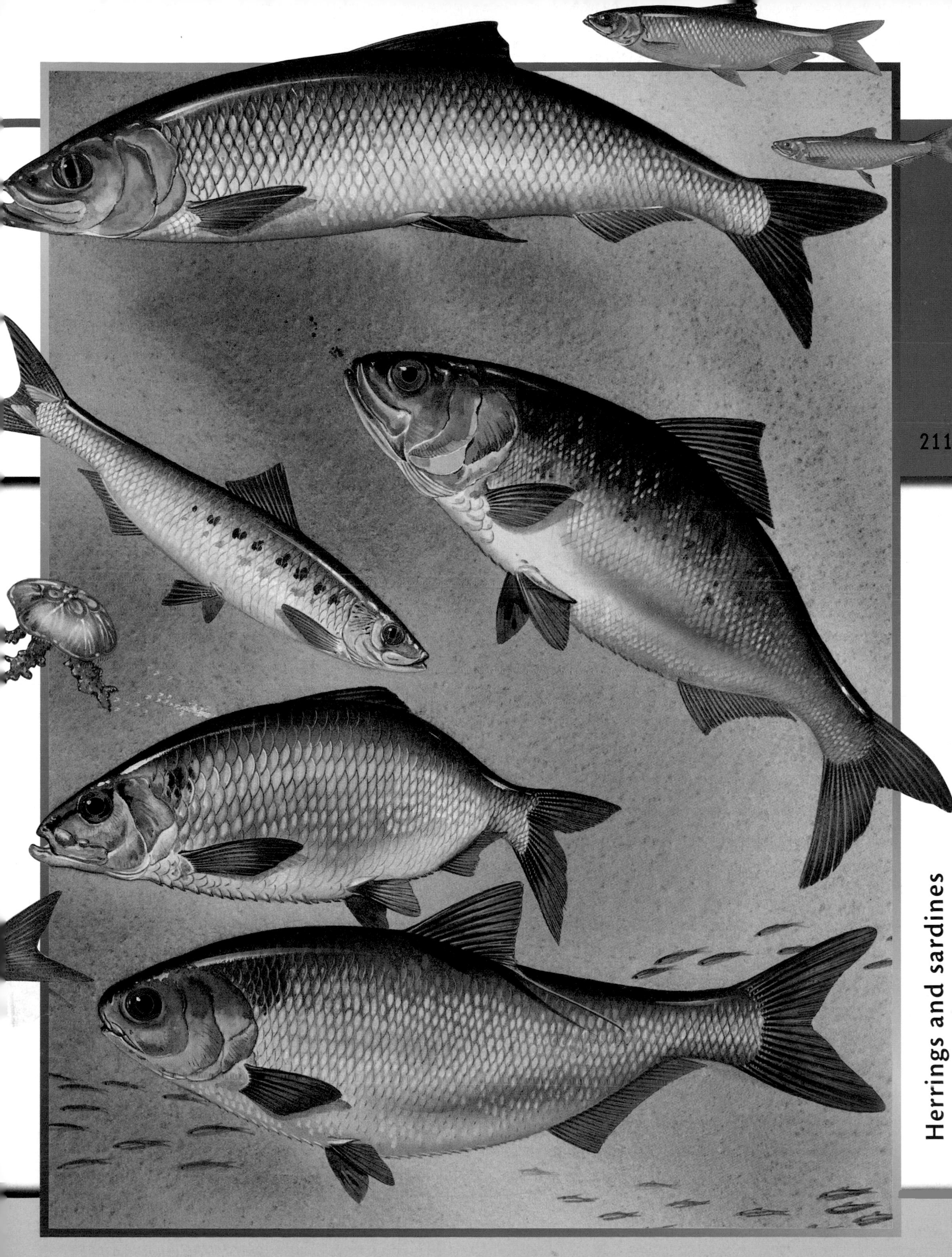

Grey shapes among the waves

- Burmeister's porpoise
- Common porpoise (harbour porpoise)
- Spectacled porpoise

Porpoises are smaller cousins of dolphins and whales. There are six kinds or species, ranging from 1.2 to about 2 m in length. They look like dark-grey, blunt-nosed dolphins. But, unlike dolphins, porpoises seldom jump out of the water. They are usually seen just breaking the surface of the sea to breathe through the blowhole (nostrils) on the top of the head. Most kinds live in small groups of up to five, often leaving one group to join another. Porpoises hunt fish and squid but also take other small animals such as prawns and krill. They find prey by eyesight and by sonar-type clicks, as in dolphins, and catch food with their small, spade-like teeth.

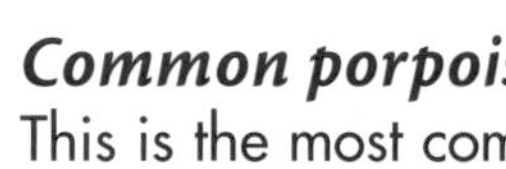

Common porpoise

This is the most common and also the most often-seen porpoise – partly because it has a very wide distribution, including the waters of the North Pacific and North Atlantic Oceans, and the Mediterranean and Black Seas. It is also because the common porpoise swims nearer shores and is not so shy of ships and busy waterways as the other species. However it is nowhere near as common as it once was, having suffered a drastic fall in numbers in the Baltic, North and Mediterranean Seas. This is probably due to pollution and lack of food (caught by fishing boats). It lives in groups of about 10 and feeds mainly on fish and cuttlefish.

Spectacled porpoise

This is one of the most attractive of the porpoises, named after the white eye markings which make it look as if it's wearing spectacles. It is also one of the largest porpoises. The males grow to an average length of 2 m, the females to about 1.8 m. Their range includes the coastal waters of South America, especially the eastern shores, also the Falkland Islands to the south-east and possibly across the ocean to New Zealand. Fish such as anchovies as well as squid probably feature in their diet.

Burmeister's porpoise

Unusual in being dark grey nearly all over, apart from a few light patches on the belly, Burmeister's is a rare and little-known type of porpoise. It is found mainly off the coasts of South America, from Peru around the southern tip of the continent and north to Uruguay. It has also been sighted around the Falkland Islands. It has a much flatter head than other porpoises and grows to about 1.6 m long and 50–60 kg in weight. Like other porpoises, Burmeister's probably eats mainly fish and squid.

The common porpoise has more than 100 teeth and can dive to depths of 90 m.

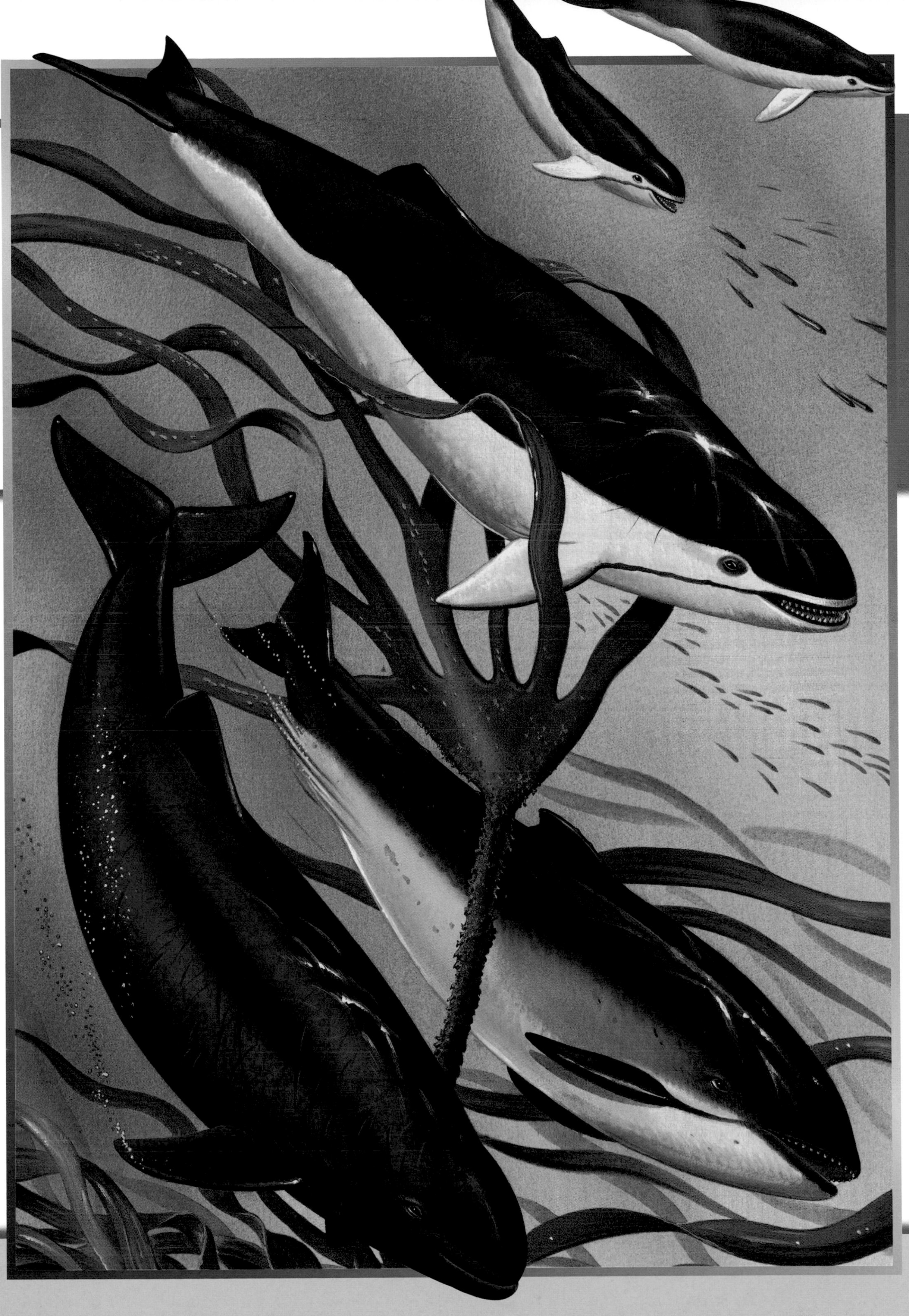

Peaceful plant-eaters of the sea

- Dugong (Indian Ocean sea cow, sea pig or sirenian)
- American manatee (West Indian or Caribbea manatee, sea cow or sirenian)

Manatees and dugongs are called 'sea cows' because they peacefully munch seaweeds and other water plants – just as cows munch grass on land. In fact these large, tubby mammals are more like hippos with flippers and live in tropical waters. The dugong has a slightly forked tail, like a whale, and lives in the Indian Ocean from the east coast of Africa across to Western Australia. The three kinds of manatees all have rounded, spoon-like tails. The West African or Senegal manatee, similar to the American manatee shown here, is found in the rivers and lakes of West Africa. The Amazon or South American manatee is smaller and also a river-dweller.

American manatee

This type of manatee is found in warm waters from Florida, USA around the Caribbean to northern South America. It lives in the sea and also swims into rivers and lakes. It can grow slightly larger than the dugong, 4.5 m long and 1500 kg in weight. Like the dugong, the manatee's front limbs are flipper-shaped. It has no back limbs at all, but its tail has become flattened for swimming. Manatees eat huge amounts of water plants – and this can come in useful. In some areas they have been captured and moved to rivers, lakes or canals which are choked with weeds. The manatees steadily chew away the weeds and clear the waterway. However manatees, like dugongs, are at risk of injury by boats. And if they become stuck in underwater fishing nets they will drown.

Dugong

The dugong grows to about 4 m long and more than 800 kg in weight – far bigger than a cow on land! It prefers shallow, sheltered coastal waters. Its favourite food is seagrass which grows down to depths of about 5 m. A female dugong does not have her first baby until the age of about 20, and she feeds it on her milk (as all mother mammals do) for about two years. So, although she may live to 50 years old, she only produces five or six young. This slow breeding is one reason why dugongs, having been hunted for their meat, are now rare in some areas. Also in today's busy coastal waters both dugongs and manatees are at risk from ship and boat propellers, hulls, waterskis and jetskis.

THE LEGEND OF THE MERMAID

Dugongs and manatees are mammals and so they breathe air. They usually come to the surface every minute or two and take a breath, although they can stay under for several minutes in an emergency. They also poke their heads above the surface to look around for danger, mates, food and other items of interest. Old-time seafaring people said that the creature's large head with its beady eyes could almost be mistaken for a person in the water. The sudden appearance of a manatee or dugong head may have led to ancient sailors' tales of mermaids. In olden times sailors spent many lonely months on the ocean. Seeing a mermaid might be the result of wishful thinking!

The manatee has the longest guts (intestines) of any animal – more than 40 m!

The greatest creatures on Earth

- Blue whale (Sibbald's rorqual or sulphur-bottomed whale)
- Bowhead (Greenland right whale)
- Fin whale
- Grey whale (shown with calf, also group or pod of greys)
- Minke whale

There are 10 kinds of great or baleen whales. They are flesh-eaters but they have no teeth. Their huge mouths contain bristly, comb-shaped plates made of a springy, plastic-like substance, baleen. The plates hang like curtains from the roof of the mouth. The whale gulps in water, closes its mouth and pushes the water out with its tongue. The baleen works like a sieve to filter out small creatures – especially shrimp-like krill.

Blue whale

The blue is not only the largest whale, but also the largest mammal, and probably the biggest creature that ever lived. It can measure more than 30 m long and weigh 160 tonnes. It also eats 4 tonnes of krill each day in the rich summer seas around Antarctica. Blues migrate towards the tropics for winter.

Bowhead

The bowhead occurs in Arctic waters, especially the Bering Sea. It has still not recovered from the days of mass whaling and it may number only a few thousand. It has massive curving jaws, a huge gaping mouth and the longest, finest-fringed baleen plates of any whale, more than 4 m in length.

Grey whale

Grey whales dive as deep as 100 m to the ocean floor and plough their mouths through the mud to scoop up shellfish, worms and other sea-bed animals. Grey whales make the longest mammal migrations, up to 15,000 km between their winter breeding grounds off the Mexican coast and their summer waters of the Bering Sea.

Minke whale

Resembling a miniature blue whale, the minke is the smallest great whale at 'only' 11 m long. It has short baleen plates and catches squid and fish.

Fin whale

The fin whale is second only to the blue whale in size, at 25 m long and 80 tonnes in weight. It is found in all the world's oceans. Fin whales feed mostly at the surface or just below and they catch small fish and seabirds as well as krill. They are also the fastest of the great whales, powering along at speeds of more than 30 km/h for many hours.

The biggest animal baby is the blue whale calf. At birth it is more than 7 m long and 2 tonnes in weight. By the age of 7 months it weighs almost 25 tonnes.

The open ocean

- Banded sea snake
- Broad-billed prion (whale bird)
- Comb-jelly (ctenophore)
- Common squid
- Commerson's dolphin
- Cuttlefish
- Dusky dolphin
- Great skua
- Kittiwake
- Mako (mako shark)
- Man'o'war jellyfish
- Nautilus
- Northern right whale
- Ocean sunfish
- Pacific white-sided dolphin
- Ridley's turtle
- Sei whale
- Tarpon
- Tiger shark
- Violet sea snail
- Yellow-bellied sea snake
- Yellowfin tuna

Endless waves

● The immense oceans are twice as large as all other global habitats combined.

Oceans and seas cover two-thirds of our world. Far out in the middle there is just wind, waves – and wildlife. Creatures seem to have endless freedom. There are no cliffs, rivers or similar physical barriers. However there are ocean currents where the water is a different temperature and salinity (amount of dissolved of salt). Also some areas are richer in food since the water has more dissolved nutrients. These features affect where many animals swim. And there is always danger on all sides. Large seabirds may swoop down to snatch fish or squid from near the surface. Sharks and other predators dash in from the sides or lurk in the gloom below.

Northern right whale
Very bulky indeed, the right whale is about 17 m long and weighs 50 tonnes. It sieves plankton from the water like other great whales.

Cuttlefish
This speedy, active hunter can change colour far faster than a chameleon. Its large eyes see prey and its two long, main, suckered tentacles grab the victim.

Tiger shark
Named after its stripes and also its aggressive hunting power, the tiger shark reaches 5 m in length. Its strong, sharp teeth can saw through turtle shells.

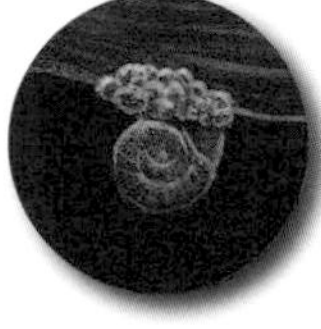

Violet sea snail
This snail produces a mass of bubbles made from its own slime, and floats upside down underneath. It eats any small bits of floating flesh it comes across.

Ridley's turtle
Smallest of the marine (sea-going) turtles, Ridley's is only about 60–70 cm long. It crunches up sea snails, shellfish and crabs in its strong jaws.

Tarpon
The tarpon is a massive hunter that can grow more than 2 m long. It lives mainly in the Atlantic Ocean, swims in shoals and hunts smaller fish.

Ocean sunfish
The sunfish is the biggest of all bony fish (as opposed to sharks), almost 4 m 'tall' and weighing 2 tonnes. Its front teeth form a bird-like beak for eating small jellyfish and comb-jellies.

Yellowfin tuna
One of the smaller tunas, the yellowfin is 'only' 2 m long. It is a fast, high-energy predator always on the hunt for squid, fish and other prey. As it gets older its fins grow even longer.

Yellow-bellied sea snake
Like land snakes, sea snakes breathe air. But they can stay underwater for more than one hour as they feed. These sea snakes occur in huge 'rafts' of many thousands.

The song of the sei whale is such a loud, low and constant hum that scientists at first thought it was the throbbing engines of a secret spy submarine!

Comb-jelly

Comb-jellies are see-through creatures similar to jellyfish. They swim by rows of tiny hairs along the body which resemble combs and wave like tiny oars. Whip-like tentacles trail below and catch small prey such as shrimps. The biggest comb-jellies are almost 100 cm long.

Pacific white-sided dolphin

This is one of the dolphins most likely to leap and 'play' in the waves made by fast ships. It lives mainly in the North Pacific and grows to about 2.2 m in length. Various fish such as herrings, sardines and anchovies make up its main food.

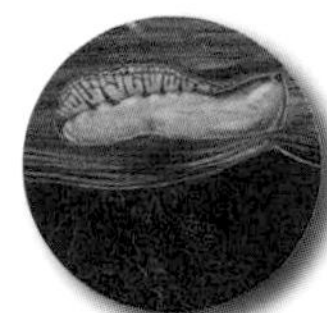

Man'o'war jellyfish

The man'o'war is not a single animal but several jellyfish-like creatures living together in a colony. One forms the sail-like, gas-filled 'float'. Those around the edge each have one long, stinging tentacle to catch prey such as fish and shrimps.

Dusky dolphin

This is a fairly short but plump dolphin at about 1.8 m long and 120 kg in weight. It also lacks the long beaky snout of typical dolphins and has a blunter nose, more like a porpoise. It feeds on fast-swimming fish and squid.

Sei whale

The sei is one of the largest whales, growing to 20 m in length. It is also probably the fastest great whale, surging along at 50 km/h for many hours. A female and male may stay together for a long time, raising one calf every two years.

Nautilus

This strange shellfish in its snail-like shell, 30 cm across, is a relative of the octopus and cuttlefish. It grabs a fish or other prey in its rings of tentacles, then rips it apart with its mouth at the centre, which has a strong, parrot-like beak.

Kittiwake

The kittiwake is a smallish, delicate gull that walks so seldom it has small legs and only three toes per foot. Most of the time it soars and dives after fish, shellfish and similar prey.

Broad-billed prion

This prion has bristle-like fringes inside its beak. They filter the water for tiny animals and other bits of floating food. This prion lives in all southern oceans.

Great skua

The sharp-beaked great skua is often called the 'pirate' among seabirds. It flies at and pecks other birds to make them drop their food, which it then steals and eats. It also takes eggs and kills and eats bird chicks.

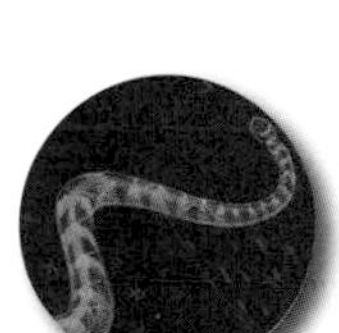

Banded sea snake

The sea snakes are so well adapted to swimming that they are floppy, weak and helpless on land. The banded sea snake eats small fish, which it kills with its incredibly poisonous bite.

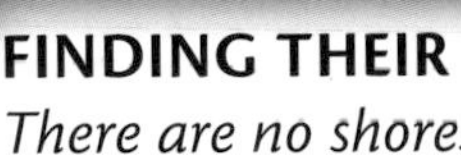

FINDING THEIR WAY

There are no shores, islands or other landmarks in the middle of the ocean. How do migrating creatures like turtles and great whales find their way? They probably steer using a combination of methods. These include the positions of the sun and moon, the directions of the currents, and the hills, valleys and contours of the sea bed where water is shallow. Some may detect the varying saltiness of sea water, and tiny variations in the Earth's magnetic field and its pull of gravity.

Kittiwakes have been seen flying over the great floating ice raft on the Arctic Ocean, within 200 km of the North Pole.

A tough life at the seaside

- Common dragonet (dragonet)
- Greater sand-eel (sand lance)
- Neon goby
- Pallid goby
- Spotted dragonet
- Tiger goby

Almost any rock pool has its small fish – mainly gobies with their slippery leathery skin, big blunt heads, strong spiny fins and tapering tails. The shore is a very tough habitat with crashing waves, rolling boulders, hot sun, driving rain, chilly winds and the rise and fall of the tides. Gobies are hardy little fish and there are more than 400 kinds around the world, living mostly along shores and in shallow seas, with some in lakes and rivers. They tend to stay on or near the bottom, using their large fins to grip seaweed and rocks. Dragonets are also shore-dwelling, bottom-living fish. Like gobies they are in the huge group called perch-like or spiny-finned fish.

Pallid goby
The pale, mottled silvery-brown of this goby is ideal camouflage on a sandy sea bed scattered with rocks and stones. Its fin spines are stiff and sharp, especially along the back (dorsal) fins. This puts off gulls, otters, bass, octopuses and other predators. The pallid is average length for a goby, about 12 cm.

Neon goby
Gobies of tropical waters tend to have brighter colours than those in colder seas, to be noticed among the corals. The neon goby often rests on its pectoral (front side) fins, as though leaning on its 'elbows'. Like most gobies it only swims in short bursts. Otherwise it remains still, watching for food or danger.

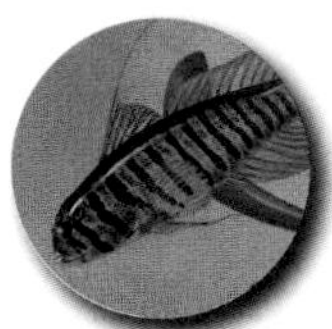

Tiger goby
Like other gobies, the tiger goby feeds on a variety of small or young water creatures such as sand shrimps, prawns, crabs and sea-snails. Its stripes conceal it among the fronds of seaweeds and sea-grasses such as eel-grass. Its thick-skinned, tough-scaled, slimy body allows it to wriggle between pebbles.

Spotted dragonet
Dragonets spend most of their time lying on or half-in the sea bed, or grubbing in sand, mud and pebbles for small animals to eat. Their eyes and gill openings are high on the head. The spotted dragonet is 15 cm long and the male has taller fins and brighter colours than the female.

Common dragonet
The male dragonet (shown opposite) is 30 cm long and a very colourful fish, with two tall back fins like brightly patterned yacht sails. The female is smaller and mainly brown, with shorter fins. Common dragonets live along the shores of the East Atlantic Ocean and Mediterranean Sea.

Greater sand-eel
Sand-eels are not proper eels but long, slim, eel-shaped members of the perch group, and close cousins of gobies. They grow to about 20 cm long and can quickly wriggle like worms into the sandy sea bed, almost out of sight. Sand-eels are common food for larger fish and birds such as puffins.

The world's smallest fish is a goby. The Philippines dwarf goby lives in a few lakes and streams on a few Philippine islands. It is this big: »≈›

Ocean soarers and divers

- American white pelican
- Australian gannet
- Frigatebird
- Red-tailed tropicbird

Many birds of the open sea have stretchy, balloon-like throat pouches which they use to scoop up fish. Pelicans are among the largest and heaviest of these ocean soarers (see also page 16). They are clumsy on land and walk with an awkward waddle. But in the air they glide without effort on their broad wings, using winds and rising air currents (thermals) to stay aloft. Gannets also soar over the sea in flocks. When they spot a shoal of fish just below the surface they fold their wings back and plunge like arrows into the water to feed. Frigatebirds and tropicbirds are graceful gliders and spend days at a time soaring over the world's warmer oceans.

Frigatebird

Frigatebirds nest on islands in tropical seas. With their long, narrow, angled-back wings and deeply forked tails, they are exceptionally graceful and skilled fliers. They can glide slowly and also beat their wings rapidly to fly against gale-force winds. Frigatebirds sometimes use their speed to chase other seabirds and make them drop their food, which the frigatebird then eats. The male has a red balloon-like pouch on his neck. He inflates this to attract the white-fronted female during breeding time.

Australian gannet

Gannets are powerful, ocean-going seabirds which usually feed in flocks. They glide and soar over the waves, watching for fish below. When the flock finds a shoal, the gannets go into a frenzy of dive-bombing as bird after bird drops into the sea, emerging with a fish in its beak. To help this plunge-dive method of feeding the gannet's body with its wings folded back is shaped like a torpedo, and its beak is long and sharply pointed. In tropical regions most types of gannets are known as boobies.

Red-tailed tropicbird

The most distinctive feature of the beautiful tropicbird is its tail streamers, formed by narrow, extra-long central tail feathers. The feathers may measure up to 50 cm in length. Tropicbirds spend most of their time on the wing over the open sea, watching for fish, squid, shrimps and similar creatures just below the surface. They catch their food by hovering and diving onto it in a flurry of spray. Tropicbirds breed on rocky cliffs and outcrops on ocean islands – the only time they come to land.

American white pelican

This pelican is a gregarious or social bird, which means it is nearly always seen in flocks with others of its kind. The flock usually rests on seashore rocks or on a sandbar. Then they take off with a whirlwind of wingbeats to settle on the sea nearby and scoop up large pouchfuls of water and fish. The pouch is made smaller to spill out the water and leave the food, which the pelican swallows. These birds are also found on inland lakes, where they make big, untidy nests from sticks, waterweeds and stones.

The Australian pelican has the longest beak of any bird – up to 47 cm.

Life on the ocean wave

- Fulmar (fulmar petrel)
- Leach's storm petrel
- Light-mantled sooty albatross
- Manx shearwater (common shearwater)
- Wandering albatross
- Wilson's storm petrel

'Tube-noses' are birds who spend almost their whole lives soaring over the open ocean. This odd name comes from the position of their nostrils (breathing openings). These are not at the base of the beak just below the eyes, as in most birds, but part-way along the beak at the end of tube-like extensions. The group includes petrels, fulmars, shearwaters and the bird with the longest wings in the world – the wandering albatross. They glide and soar in all weathers, even over mountainous waves in howling gales. These birds feed by swooping down and snatching fish, squid and other food from the surface, using their long, sharp, down-hooked beaks.

Light-mantled sooty albatross

There are 14 kinds of albatrosses. This is one of the smallest with wings about 1.5 m across. It lives in the far south around Antarctica. Unlike other albatrosses, which make untidy nests or none at all, the female and male build a tidy, bowl-shaped nest from bits of plants. They rear a single chick.

Wandering albatross

You can watch a wandering albatross for hours and never see it flap its wings. It glides quite low, usually less than 20 m above the waves, where rising winds keep it aloft. Albatrosses learned the trick of following ships for thrown-away food scraps, so they feature in many sailors' tales.

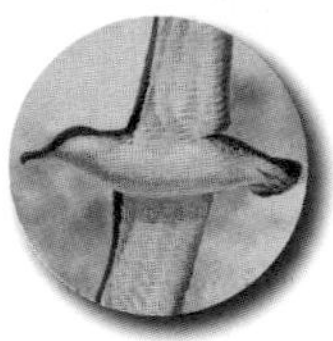

Manx shearwater

This common seabird is found around the Pacific and Atlantic Oceans and in the Mediterranean Sea. It wanders widely over the ocean for weeks on each feeding trip, covering up to 500 km each day. Then it returns to its breeding colony where thousands of these shearwaters nest in burrows.

Leach's storm petrel

Storm petrels are among the smallest seabirds, many measuring less than 20 cm from beak to tail. Yet they can fly even in the worst storms, or sit on the sea as giant waves crash around them. Leach's storm petrel lives in northern oceans and feeds by swooping near to the water's surface.

Wilson's storm petrel

Storm petrels of southern oceans, like Wilson's, often feed by swooping down and then 'stepping' on the ocean surface. They seem to run or hop along as they stab with their beaks for small food items like krill. These petrels only come to land during the breeding season.

Fulmar

This large, strong, gull-like petrel has become common because it feeds on scraps from fishing boats and bullies other birds away from this food. Like other tube-noses the fulmar keeps a store of food in its stomach, in the form of thick oil. If a predator comes near the bird vomits up the oil as a foul-smelling spray!

The wings of the wandering albatross can measure 3.2 m from tip to tip.
Wilson's storm petrel is one of the five commonest birds in the world.

Mysterious whales of the deep

- Baird's beaked whale (northern four-toothed whale)
- Cuvier's beaked whale (goose-beaked whale)
- Sowerby's beaked whale (North Sea beaked whale)
- Northern bottlenose whale

There are 18 different kinds or species of beaked whales, named after their bird-like beaked noses. Most are between 5 and 10 m long and look like a combination of great whale and dolphin. But they are mysterious and seldom-seen creatures and little is known about their lives. They spend most of their time far below the surface, diving to enormous depths. They probably swim close to the sea bed and follow underwater hills and valleys as they hunt squid, fish and similar victims. Beaked whales usually have marks and scratches over their backs. These may be the results of fights with their prey, or with breeding rivals of their own species.

Cuvier's beaked whale

In this beaked whale the nose 'beak' is quite short and the forehead is less bulbous than in other beaked whales. (The bulging foreheads of beaked whales led to their nickname of 'barrel-heads'.) Cuvier's beaked whale ranges widely through the world's seas. It reaches 7 m in length and 6 tonnes in weight. The white patches on its body are often caused by parasites such as fish-lice and barnacles growing on its skin. It is named after Baron Georges Cuvier, the great French scientist and fossil expert from the 1800s.

Baird's beaked whale

Sometimes known as the northern giant bottlenose, this alternative name gives a good idea of its appearance – similar to a giant version of the familiar bottlenosed dolphin. This is the largest beaked whale, growing to 12 m long and 13 tonnes in weight, although females are slightly larger than males. These elusive whales live in the waters of the North Pacific Ocean, descending more than 1000 m when hunting. They have just four teeth, each about 6–7 cm long, two on either side near the front of the lower jaw.

Northern bottlenose whale

This whale has the largest, most bulging forehead of the whole group. It lives in the deep waters of the North Atlantic, usually travelling about in small herds like other beaked whales. It can dive to 1000 m or more and stay underwater for two hours! Northern bottlenose whales migrate north towards the Arctic in spring and back to warmer waters in autumn. They are about 9–10 m long and 10 tonnes in weight when fully grown. Like most beaked whales they are long-lived, reaching perhaps 40 years of age.

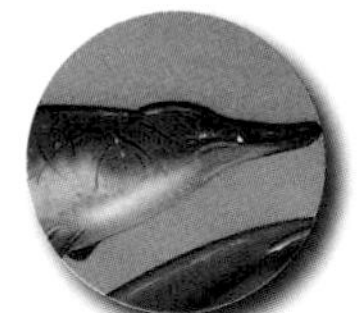

Sowerby's beaked whale

This small beaked whale grows to 5 m long and is found mainly in the North Atlantic Ocean. It is occasionally stranded on European shores. Sowerby's beaked whale is a member of a sub-group of beaked whales which have just two teeth. These are quite large and grow towards the front of the lower jaw. All of these whales have them but the teeth only erupt, or grow out of the gum, in older individuals. Then the teeth can be seen protruding from the mouth. This whale dives to 2000 m and feeds mainly on squid.

Living specimens of one type of beaked whale, Shepherd's beaked whale, have only been seen about 10 times.

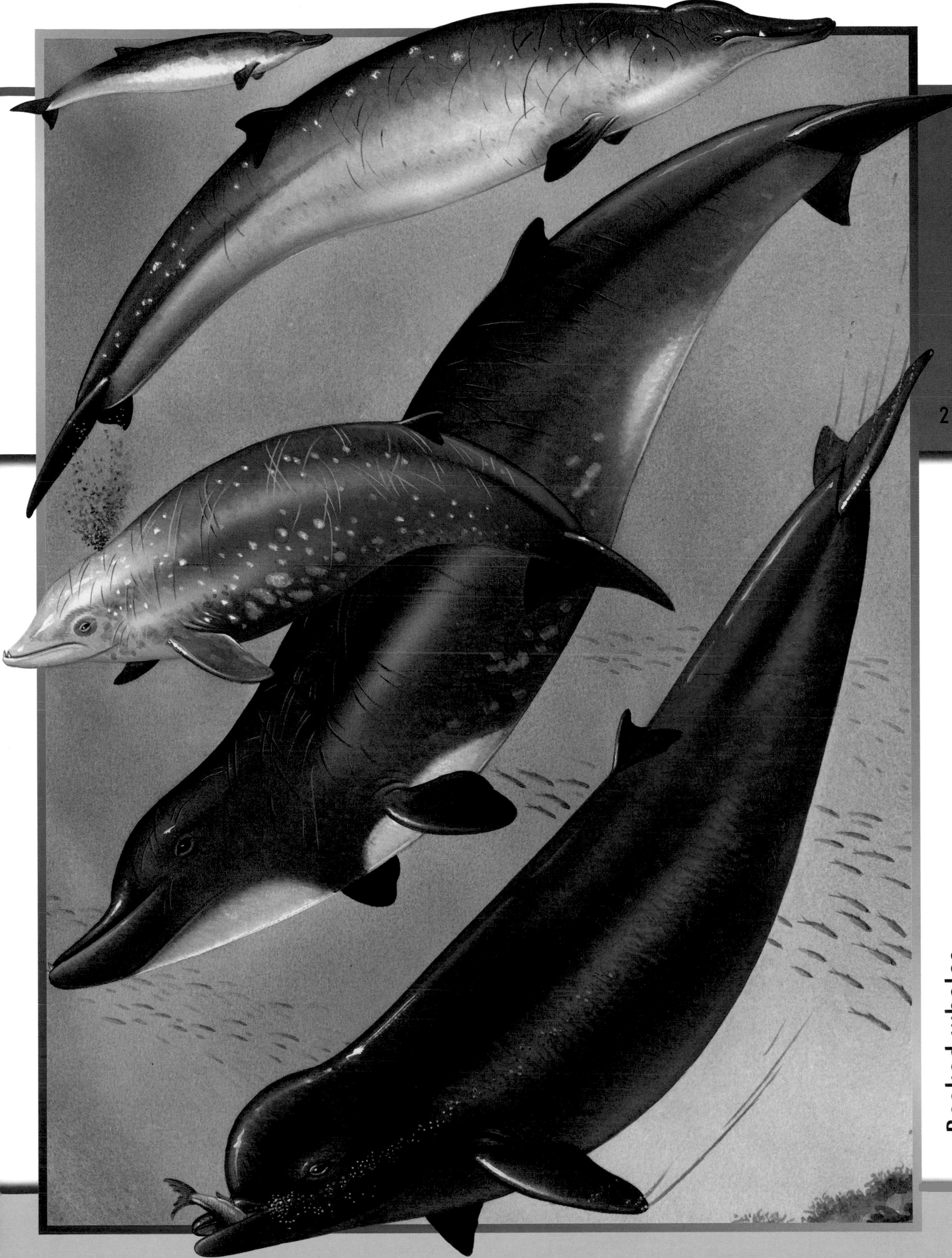

Flying fins and needle noses

- Atlantic flyingfish (four-finned flyingfish)
- Ballyhoo
- Flyingfish (two-finned flyingfish)
- Garfish
- Houndfish

Why do flyingfish fly? They don't – at least, not true flapping flight. They swim fast just under the water's surface, then leap above to glide on their large outstretched fins. This is usually to avoid predators such as tuna and dolphinfish which are chasing them just below. Flyingfish belong to a large group of more than 800 kinds of fish which also includes toothcarps, silversides, killifish, needlefish and halfbeaks like the ballyhoo. Halfbeaks are named after the oddly shortened upper jaw. Needlefish such as the garfish have very long, thin, sharp jaws. All of these fish are active predators, chasing and eating smaller sea creatures.

Garfish
The garfish is quite different from the gars or gar-pikes of North America. It is found in the North Atlantic Ocean and the Mediterranean and Black Seas, where it speeds through the surface waters after prey. Garfish grow to about 90 cm long and can skitter just above the surface by swishing their tails hard.

Flyingfish
Slightly smaller than the Atlantic flyingfish, at about 30 cm in length, this type has only two enlarged fins. It cannot glide quite so far or with such skilled control. Flyingfish in a shoal which are threatened by a pack of larger predators sometimes leap from the water in their hundreds, like a flock of silvery, scaly birds.

Ballyhoo
The ballyhoo lives mainly in the West Atlantic, in shallow water where it eats small creatures as well as bits of plants such as sea-grass. Its strange mouth may help it scoop food from the surface. Like garfish, the ballyhoo and other halfbeaks can skim or skitter across the surface, with the front part of the body in the air.

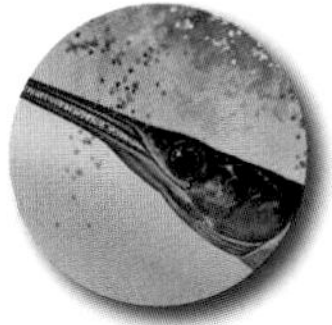

Houndfish
The largest type of needlefish, almost 1.5 m long, swims fast and flicks its head sideways in a flash to seize a fish in its long, sharp-toothed jaws. Then, almost like a heron or similar bird, it tosses the fish around to swallow it head-first. Houndfish skimming the surface have flipped into small boats and given severe bites!

Atlantic flyingfish
The very large pectoral fins work like wings so the flyingfish can glide for many seconds through the air. The Atlantic flyingfish also has slightly enlarged pelvic fins, behind and below the main ones, for better control. It reaches a take-off speed of about 50 km/h and can stay airborne by thrashing its tail to and fro 50 times each second, with the longer, lower part just dipped into the water. Back under the water the flyingfish folds up its large fins like fans and holds them against the sides of its body so they do not slow it down. This flyingfish reaches 40 cm in length and lives in shoals. Sometimes a stormy gust of wind lifts several of them so high during their glides that they land on the deck of a ship.

A good glide for a flyingfish covers about 100 m in 10 seconds, cruising at an average height of 1.5 m above the surface.

The most feared fish in the sea

- Basking shark
- Blue shark
- Great white shark (white pointer, white death or maneater)
- Sandbar shark
- Starry smooth-hound (starry hound-shark)
- Whale shark

The word 'shark' brings an image of the supreme predator – a sleek, fast, powerful hunting fish, with a wide mouth full of razor-sharp teeth, and incredible senses able to detect faint traces of blood in the water from 5 km away. This is true for many sharks such as the blue, mako, tiger and the dreaded great white. However the biggest sharks of all, the whale and basking sharks, are peaceful plankton-feeders and almost harmless. All 360 or so different kinds of sharks have skeletons made of cartilage (gristle), like rays, rather than made of bone as in other fish.

Great white shark At some 6 m in length, the great white is the largest predatory shark. It swims in warmer waters around the world. It is not actually white but grey or brown on top.

Starry smooth-hound Hound sharks and their smaller cousins, dogfish, are named from their habit of prowling in small shoals or 'packs' for bottom-living prey like crabs, prawns, worms and shellfish. This shark grows to 1.5 m.

Whale shark The world's largest fish, the whale shark takes in huge mouthfuls of water and its specialized gills filter out small floating animals and plants.

Basking shark Almost as large as the whale shark, at about 11 m long and 3–5 tonnes in weight, this giant fish is named because it often lazes in the sunlit surface waters. Like the whale shark it filters plankton from the water using comb-like bristles or rakers on its gills. It sometimes gathers in large shoals.

Blue shark The super-streamlined blue lives in warm seas and is 3.7 m in length. It has very long side fins and eats surface fish like herring and mackerel.

Sandbar shark Light brown to match the sandy bottom, this shark is a shallow-water predator in tropical and subtropical regions. It is a fast swimmer but sometimes lies still on the sea bed, only its gill muscles working so it can obtain oxygen from the water.

The whale shark grows to a length of 17 m or more and a weight of about 10 tonnes.

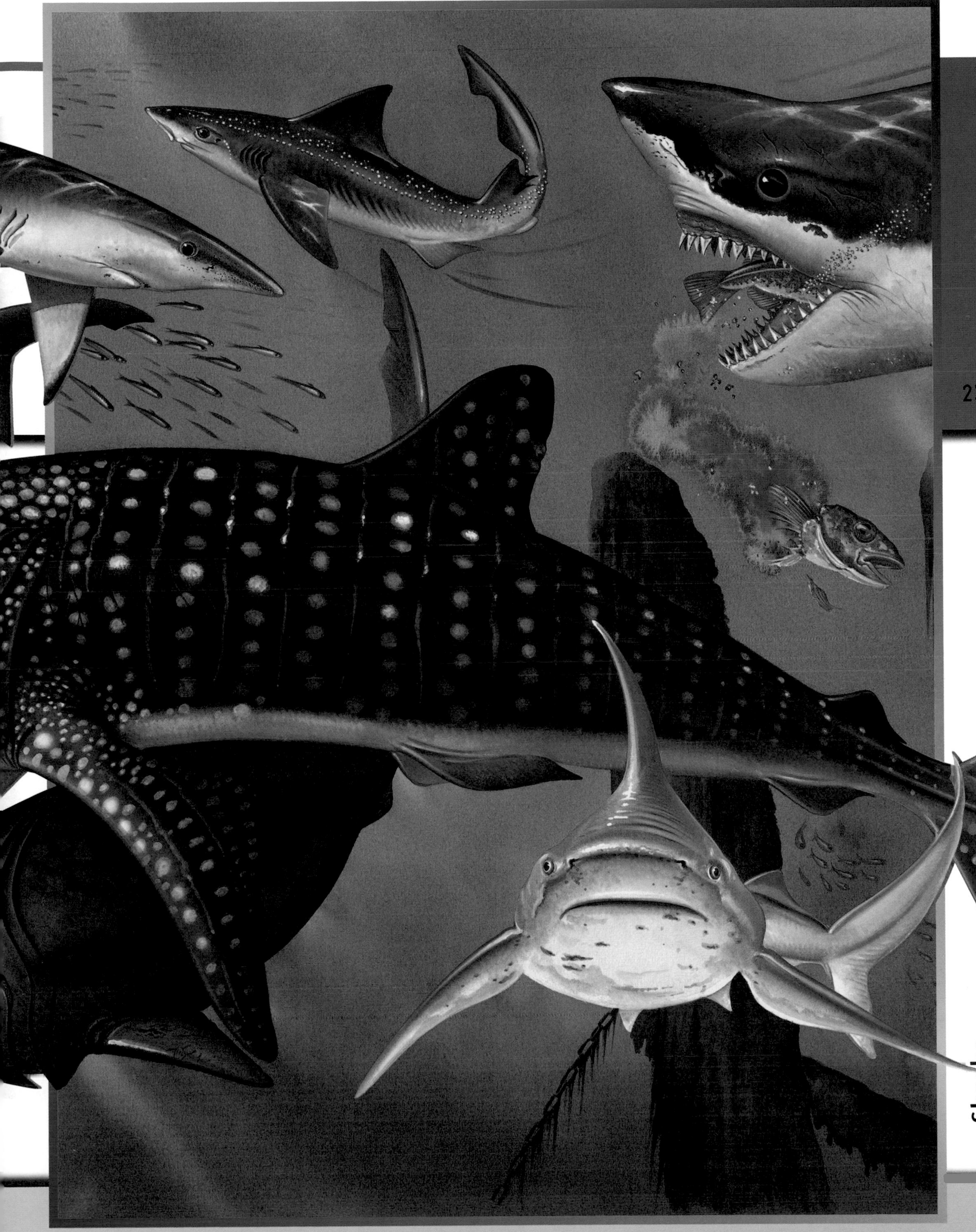

Giant hunters of the depths

- Dwarf sperm whale (Owen's dwarf sperm whale, rat-porpoise)
- Pygmy sperm whale (small, lesser or short-headed sperm whale)
- Sperm whale (cachalot, spermacet or spermaceti whale)
- Squid

Sperm whales spend much of their time in the blackness of the deep ocean, hunting their prey by the clicks and squeals of their sonar. They regularly dive more than 1000 m below the surface and stay under for an hour as they hunt near the ocean floor for fish, squid, crabs and similar food. The sperm whale's enormous head makes up one-third of its whole body length. It contains a waxy or oily substance known as spermaceti which was once used to make candles, cosmetics, creams and very high-quality lubricating oil. Sadly, hunting for this and for other body parts, like the flesh and blubber, has made sperm whales rare.

Sperm whale

The mighty sperm whale is more than 20 m long and 40 tonnes in weight, making it the largest toothed whale – and by far the largest predator or meat-eater on Earth. When whales surface to breathe they can be recognized by the pattern of moist, steamy air they snort out through their blowholes. In the sperm whale this 'blow' is angled forwards and to the left.

The massive bulging head of this whale is filled mainly by the spermaceti organ, which has layers of oily wax. These act as a kind of sound-lens to focus the whale's huge grunts of low-power sound which can stun its prey. The spermaceti organ also works as a buoyancy device. As the whale descends it becomes denser or heavier, making the whale less buoyant and so the dive is easier.

Dwarf sperm whale

Similar in shape to a porpoise, without the usual sperm whale's bulging forehead, the dwarf sperm whale reaches about 2.5 m in length. It is found most often in warmer seas, especially off the coasts of South Africa, India and Australia, and rarely in the open ocean like the pygmy sperm whale.

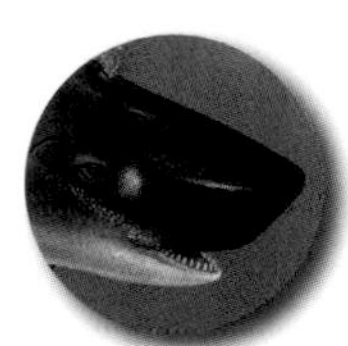

Pygmy sperm whale

Pygmy sperm whales grow to about 3.4 m long and 500 kg (half a tonne) in weight. They have a varied diet including shrimps, crabs, fish, octopus and cuttlefish as well as plenty of the usual squid. They catch their food at depths of about 100 m, in coastal waters and far out at sea.

BATTLE OF THE TITANS

Sperm whales sometimes plunge down 2000 m or more for two hours or longer. After returning to the surface for a couple of minutes to breathe they descend again into the depths. Sometimes a group of sperm whales dive together and work as a team to round up their prey. Like dolphins, they find food by sonar or echolocation, making squeaks, clicks and grunts that bounce off nearby objects. A sperm whale has up to 50 cone-shaped teeth in its slender lower jaw but no teeth showing in the upper jaw. Its skin is scarred by marks from the suckers and the hard, horny, beak-like mouth of its deadly enemy. This is also its biggest prey, with a 6-m body and 10-m tentacles – the giant squid.

Ambergris was once used to make the delicate scents of very expensive perfumes. Which is odd because this substance comes from inside the guts of sperm whales!

Fish with big appetites

- Atlantic mackerel (common or European mackerel)
- Bluefin tuna (bluefin tunny)
- King mackerel (cavalla or kingfish)
- Skipjack (skipjack tuna)
- Spanish mackerel

Since prehistoric times people have caught and eaten tunas and mackerels. These fast-swimming members of the vast perch-like (spiny-finned) fish group are close relatives of marlins and swordfish. To swim so fast, more than 50 km/h, tunas have huge muscles along the body and a slim, curved tail. The muscles make up more than two-thirds of the fish's whole weight, which is why tunas are such valuable food fish. A tuna also has a spiny front dorsal (upper back) fin, a softer rear dorsal fin, and small finlets along its back to the tail.

Bluefin tuna
This massive fish can grow to more than 4 m in length and 700 kg in weight. It lives in all seas and oceans but moves around in shoals with the seasons. Sometimes bluefins stay near the surface, chasing fish such as herring and mackerel. Sometimes they stay in deep water and use their big eyes to hunt squid.

Skipjack
A small tuna about 100 cm long, the skipjack is found in great numbers in the Pacific Ocean, where shoals can be as large as 40,000. Vast curtain-like nets are hung in the sea to catch these tunas. However the nets trap seabirds, seals, porpoises, dolphins and similar air-breathing animals – which soon drown.

HOT-BLOODED FISH
Fish are cold-blooded, with a body temperature the same as the water around them – except for tunas. Their muscles are so large and so active that they produce lots of heat, which keeps the tuna 'warm' in the cold sea. Warm muscles work even better, which is why tunas can swim so far so fast.

Atlantic mackerel
Mackerels are smaller relations of tunas. Atlantic mackerels are only 40 cm long. However they swim in vast shoals and have been caught for centuries as food fish. Like their huge relatives they are fast-moving, hungry predators. However during winter they lie still near the sea bed and hardly eat.

King mackerel
This strong, fierce predator has sharp, blade-shaped teeth and may prey on its smaller mackerel namesakes. The pattern along its sides varies from one fish to another and is probably for camouflage in the shadowy water. This pattern may form mainly stripes or be more broken up into spots or marble-like swirls.

Spanish mackerel
These mackerels live mainly in the central region of the Atlantic Ocean, especially in the waters off Europe and North Africa. They are larger than common mackerel at about 4–5 kg in weight. The finlets in front of the tail, as in other mackerels and tunas, help with speedy swimming.

The largest king mackerels grow to 1.5 m long and about 45 kg in weight. This is 100 times the weight of the common or Atlantic mackerel.

Swords, sails and speed

- Blue marlin
- Sailfish
- Swordfish (broadbill)
- Wahoo
- White marlin

Swordfish, sailfish and marlins are all billfish – some of the most spectacular and exciting fish in the ocean. They are named after their long, pointed snouts or bills ('beaks'). They are big, powerful, extra-streamlined predators of smaller fish, squid, seabirds and other ocean creatures. These fish have the tails typical of the fastest swimmers. The tail is thin and stiff, shaped into a slim curve or crescent, with a narrow stock or base where it joins the body. The fish does not swim by long, snake-like waves of the body. Instead its whole body shakes or vibrates to wave the tail from side to side, many times each second, in a rapid frenzy of surging power.

Wahoo

The wahoo is not a billfish but it is a close relative. Like billfish and its even closer cousins, the tunas, it belongs to the huge group called perch-like or spiny-finned fish. Wahoos reach about 2 m in length and are found in all tropical seas. They can get up speed faster than any other fish, from 0 to 70 km/h in 2 seconds.

Sailfish

Probably the world's fastest fish, the sailfish has a long, tall dorsal (back) fin. At top speed it folds this and its other fins against the body to improve streamlining. At breeding time the female sailfish releases more than 5 million tiny eggs into the water. Only a few survive and grow to adult size, 3.5 m, in 6–7 years.

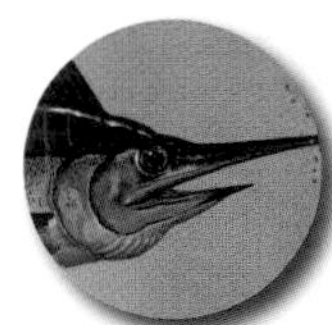

Blue marlin

Like other billfish the huge blue marlin lives and hunts mainly in the upper surface waters. These marlins regularly reach 3.5 m in length and weigh 200 kg, and a few grow even larger, over 500 kg. Marlins are much prized by anglers because of their strength, stamina and amazing leaps from the water.

White marlin

Most billfish like the white marlin make seasonal migrations. They move north or south in summer and return to the tropics in winter. The white is the smallest of the four kinds of marlin, weighing about 70–80 kg. It lives in the warmer regions of the Atlantic Ocean. (Largest is the black marlin at 700-plus kg.)

Swordfish

The swordfish may weigh 650 kg and be almost 5 m long. But about one-quarter of this is its 'sword' formed by the long upper jaw. Like other billfish it catches prey by speed and power. But how? It's very unlikely that the swordfish stabs them. More probably it swims at speed through a shoal of smaller fish, waving its head from side to side to slash through the water at the prey. The bill may actually hit victims. But it also creates powerful ripples or shock waves in the water which stun and daze them. The swordfish then returns quickly and swallows its meal. The sword grows gradually – young swordfish have much shorter noses in proportion to their bodies. One swordfish drove its nose 55 cm into the timbers of a wooden ship.

Measuring the top speed of big, fast ocean fish is very difficult. But the sailfish can probably reach speeds of about 100 km/h.

Snakes of the sea

Eels look like snakes but they are really fish with long, wriggly bodies. There are more than 600 kinds (species) and they live all over the world, mainly in warmer seas and oceans. Three of the normal fish fins – the dorsal (back), caudal (tail) and anal (underside) – join to form one long flap that wraps around the body. There are also the usual two pectoral fins on the front sides of the body, as in other fish. Eels are mostly lurking, sharp-toothed predators. They can wiggle their long, thin bodies to hide in cracks in the rocks, among coral or in holes in the sea bed. They catch mainly smaller animals such as fish, prawns, shrimps, sea-snails and worms.

Conger

Divers who mend harbour walls or explore sunken shipwrecks are always wary of the conger. It hides away, camouflaged by its dull grey-blue colour, with just its head poking out. It peers through the water with its large eyes, ready to seize any passing prey in its strong jaws. This eel can grow to almost 3 m long and easily fit a human hand into its mouth. Like most eels it travels into deep water to breed, producing small, leaf-shaped larvae or young (as explained below). These gradually change shape as they grow.

European eel

One of the world's most common fish, this eel lives almost everywhere in Europe, North Africa and East Asia, including coasts, rivers, lakes and ditches. Females grow to about 100 cm long but males are barely half this size. These eels eat almost anything and survive out of water for several hours. They even wriggle across fields to reach new, land-locked reservoirs and canals. European eels swim to the West Atlantic to breed. The small, leaf-shaped, see-through young drift back over 3–4 years with the ocean currents.

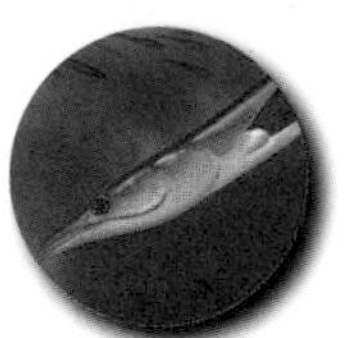

Pink snake eel

There are more than 200 kinds of snake eels. They live mainly in shallow tropical oceans and some have bright colours. But many kinds, such as the pink snake eel, are more like worms than snakes. They have very long and bendy bodies, and leathery skin which lacks scales. They dig burrows by wriggling tail-first into the mud or sand on the ocean bottom. Then each snake eel hides in its tunnel with just the front of its head sticking out, to watch and smell the water for victims.

Green moray

Divers' tales about moray eels say that their bites are poisonous, and once they have closed their teeth on something they never let go. Certainly morays are very fierce and always ready to snap, rather than slither away like most other eels. But some tales are not entirely true. A moray soon lets go if it is hauled above the water. And the bite is not poisonous, although it may go septic (bad) from germs on the moray's teeth. The green moray grows to about 2 m in length and hides among seaweed and eel-grass.

Moray eels are named after a nobleman of Ancient Rome, Lucinius Muraena. To show his power and wealth he kept these eels in tanks – more than 6000 of them!

In the warm midwater gloom

- Dealfish
- John Dory
- Lantern-eye
- Long-jawed squirrelfish
- Oarfish
- Opah

The gloomy midwater zone covers depths of about 100 m to 1000 m, depending on the clarity of the sea water – below this it's nearly or completely pitch black. In the tropical and subtropical oceans many little-known, large-eyed, colourful fish swim in the dimness, often sporting long or spiky fins. The types shown here belong to the tongue-twisting zeiform, lampridiform and beryciform groups. More simply they are all predators, hunting mainly smaller fish, prawns, shrimps, squid and other victims. Some, like the John Dory and the lantern-eye, spread into shallow waters. Others range into colder regions, like the dealfish off Iceland.

Dealfish

No other fish has such an unusual fan-like tail, pointing upwards as though joined to the body at right angles compared to other fish. There is also a distinctive red fin along almost the entire back. Dealfish grow to 2.5 m in length and dwell in the eastern North Atlantic. They feed on small squid and fish.

Oarfish

At 7 m from its snout to the end of the tapering tail, the eel-like oarfish is one of the longest of all fish. However it is far from bulky, being shaped like a strap or ribbon and only 10 cm wide. The tall crest is formed by the first few spines of the dorsal fin, which runs virtually the length of the body.

John Dory

Valued as a food fish, the John Dory has a tall but thin body when seen head-on. Like the other fish shown here it is equipped with a protrusible mouth which opens forward like a wide tube to engulf victims. The John Dory is 65 cm long and found in the Mediterranean Sea and eastern Atlantic Ocean.

Long-jawed squirrelfish

The 60-cm-long squirrelfish has large eyes, not to see in the faint light of midwater, but to hunt on shallow reefs at night. It inhabits rocky inshore areas of the western Atlantic and Gulf of Mexico. The vicious rear-pointing spine on the gill cover at each side of the head helps to deter enemies.

Lantern-eye

Most fish with light-producing organs live in deep water. But the lantern-eye, 30 cm long, favours shallows around Southeast Asia. The fish is named from the curved patch below each eye. This appears white by day. However at night it glows lantern-bright and even flashes on and off when the fish is agitated.

Opah

This bright, bulky animal, also known as the moonfish, is found in midwaters around the world. Strong and heavy, up to 1.5 m long and 75 kg in weight, it lacks teeth and looks far from speedy or agile. Yet the opah still manages to catch fast-swimming prey such as squid and whiting in its protruding, thick-lipped mouth.

At about 7 m, the oarfish is the longest bony fish. (Some sharks are longer but they have skeletons of cartilage or gristle, not bone.)

Big smiles and clever tricks

- Bottlenose dolphin (bottle-nosed dolphin)
- Common dolphin (saddleback dolphin)
- Fraser's dolphin (short-snout dolphin)
- Risso's dolphin (grampus dolphin)
- Southern right whale dolphin
- Spinner dolphin
- Spotted dolphin

Some people are lucky enough to meet real dolphins swimming free in the sea. These marvellous creatures, which are smaller cousins of whales, are warm-blooded mammals like us. But they have no fur, their arms are shaped like flippers, they have no legs either, and their tails have two broad side flukes. All 32 kinds of dolphin hunt fish, squid and similar sea animals. They take in air through their nostrils, which form the blowhole on top of the head, so they must surface every minute or two to breathe. Dolphins are fast and agile in the water. They naturally leap, somersault and spin for no obvious reason. Perhaps they are having fun?

Bottlenose dolphin
This is the type of dolphin usually seen at sea life centres. It lives in all warmer oceans and grows to about 3.5 m long. Dolphins are intelligent animals. They learn new tasks rapidly and in captivity they even invent their own tricks to play on their keepers.

Fraser's dolphin
This average-sized dolphin is about 2.3 m long and 85 kg in weight. All dolphins make clicks and squeals to find their way by echolocation and to keep in touch with their group.

Risso's dolphin
Risso's grows to 4 m long and weighs 370 kg. Its skin scars probably come from fights with its own kind at breeding time.

Spinner dolphin
This is one of the most acrobatic dolphins. It jumps high out of the water and spins around like a top. It lives in the open ocean and eats mainly fish.

Common dolphin
For centuries artists have painted and sculpted these dolphins, which have very variable markings and colours. They live worldwide.

Spotted dolphin
The speckles of this dolphin may help to camouflage it in dappled surface waters. However dolphins have few enemies, mainly large fish like sharks.

The smallest dolphin is probably Heaviside's dolphin of the waters around South Africa. It is about 1.2 m long and weighs only 40 kg. The largest dolphin is the killer whale.

Dangerous or just curious?

- Barbu
- Blue bobo
- Great barracuda (picuda or becuna)
- Southern sennet (Southern barracuda)
- Striped mullet

If all 'big fish' stories were true, barracudas would be as dangerous as those other greatly feared fish, sharks. But this applies only to one of the 18 kinds of barracudas – the great barracuda. It is a big, fierce hunting fish with long, sharp teeth. There are records of it following and even biting divers. However some tales of its savage attacks may be larger than life. Barracudas belong to the vast group of perch-like or spiny-finned fish. So do mullets. Most fish are either saltwater or freshwater. But mullets are unusual. They can swim from the salty sea into the brackish (part-salty) water of estuaries, where fresh water from a river flows into the ocean.

Barbu

The barbu is a member of the fish group known as threadfins or tasselfins. The front spine-like parts or rays of the pectoral fins, at the front sides of the body, are long and thin like lengths of string. They are sensitive and used for feeling, both by touching objects and by detecting ripples and currents.

Great barracuda

The great barracuda is well known as a fish with great curiosity. It swims near divers and follows them, watching their movements and actions. This quiet menace is one reason for the barracuda's fearsome reputation. Some attacks may be the result of barracudas being provoked or surprised by divers or swimmers. But there are also tales of sudden bites for no apparent reason. Another strange feature is that barracudas may be very aggressive in one area, yet shy and peaceful in another. They live in most warmer waters but are often met in the Caribbean and West Atlantic Ocean. They form shoals around shores and reefs when young. The large adults dwell alone, mainly in deeper waters.

Blue bobo

The bobo is a mainly bottom-dwelling type of threadfin. It lives in the warmer southern waters of the Indian and Pacific Oceans and grows to 60–80 cm long. It noses in the mud for food with its 'overhung' snout, the upper jaw being longer and more protruding than the lower jaw (as in the anchovies).

Striped mullet

The striped mullet reaches 90 cm in length and lives in warmer coastal regions around the world as well as out in the open ocean. It feeds by sucking mud and sand into its mouth and filtering out tiny edible bits using its specialized gills. The bits are swallowed and ground up in its muscular stomach-like gizzard.

Southern sennet

There are several types of smaller barracudas or sennets living in the Indian, Pacific and Atlantic Oceans – mainly in the warmer waters to the south. Some grow to about 1.6 m in length. They are powerful, sleek, sharp-toothed predators and live similar lifestyles to the great barracuda.

Great barracudas probably grow to about 2 m long, although some people claim they reach much larger sizes.

Wings underwater

- Atlantic guitarfish
- Atlantic manta (manta ray, devil ray or devil fish)
- Blue stingray (blue-spotted stingray)
- Common skate (European skate)
- Spotted eagle ray

Rays and skates are flat-bodied cousins of sharks, with skeletons made of cartilage (gristle) rather than bone. Rays may look like flatfish such as plaice but they are quite different. A flatfish rests and swims on its side while a ray rests on its true belly or underside. However both types of fish are flat for the same reason – living on the sea bed. Rays swim by rippling or flapping their 'wings' which are very large, fleshy pectoral fins. There are about 320 members of the ray group including mantas, skates, sawfish and guitarfish. Nearly all are predators. They eat mainly shellfish, crabs, worms and similar animals which they grub up from sea-bed mud.

Blue stingray

Stingrays can indeed sting, using the sharp spine part of the way along the thin tail. The venom (poison) is extremely painful, even deadly. The blue stingray grows to 1.5 m long. It is especially dangerous since it likes shallow water and hides in the sand of holiday beaches around the Indian and Pacific Oceans.

Atlantic manta

Unlike most rays, which spend hours lying camouflaged on the ocean bottom, mantas spend much of their time swimming near the surface. The Atlantic manta is a sizeable fish but smaller than its huge Pacific relative. The name 'manta' comes from the dark colour and shape of the fish, which is like a cloak or mantle. The alternative name of devil ray comes from the two fleshy flaps on the head, which look like horns. The flaps scoop water into the mouth as the manta swims by powerfully beating its vast 'wings'. As the water passes through the mouth, small creatures of the plankton are filtered out by comb-like parts on the gills. The manta may also swallow the occasional larger fish, squid, shrimp or prawn.

Spotted eagle ray

Eagle rays have especially large wing-like pectoral fins. Seen from the front they resemble the eagles of the air as they swoop gracefully at speed above the sea bed. As in several kinds of rays, the female's eggs hatch and grow inside her body and she gives birth to fully-formed young.

Common skate

This skate is a massive, powerful fish some 2.7 m long, almost 2 m wide and more than 110 kg in weight. It preys on all kinds of victims at all levels in the water – not only smaller surface fish like herrings but also other rays, small sharks like the dogfish, and also flatfish, crabs and lobsters on the sea bed.

Atlantic guitarfish

Guitarfish look like a mixture of ray and shark with longer, slimmer bodies than most other rays. The Atlantic guitarfish is found in warmer waters along the east coast of North America. It reaches about 70 cm in length and crunches up crabs and shellfish with its large, flat crushing teeth.

The Pacific manta is the largest ray with a length of about 5 m and a 'wingspan' of almost 7 m.

The bottom of the sea

- Dragonfish
- Gulper eel (swallower eel)
- Long-rod anglerfish
- Sloane's viperfish
- Snaggletooth
- Tassel-chinned anglerfish
- Tripod fish

The world's biggest habitat is also its most mysterious. The ocean depths are endlessly black and cold. Fish and other animals live on the 'rain' of rotting debris floating down from above – or eat each other.

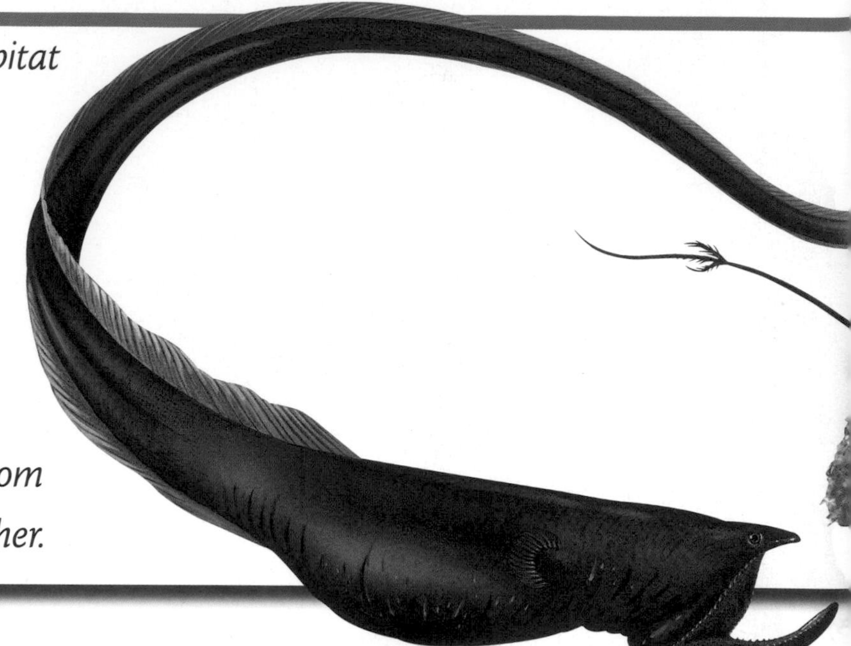

Gulper eel
Prey is scarce in the vast blackness of the deep ocean, so fish like the gulper have large mouths to grab whatever they can. This eel is a relative giant of the depths at 60 cm long.

Snaggletooth
Several deep-sea fish have rows of glowing or bioluminescent spots along their sides. These may signal to others of their kind at mating time.

Tassel-chinned anglerfish
Hardly larger than your thumb, this anglerfish has extraordinary fleshy tassels on its chin that resemble seaweed.

Tripod fish
The tripod fish 'walks' along the soft mud of the sea bed on the long spines of its two lower side fins (pelvics) and lower tail. It probably eats small shrimps and similar shellfish.

Long-rod anglerfish
The body of this anglerfish is only 15 cm long but its bendy, whip-like 'fishing rod' can be more than 20 cm in length.

Dragonfish
Like snaggletooths, dragonfish are predators of smaller creatures. They rise nearer the surface at night to follow their prey such as very young squid.

Sloane's viperfish
The first spine or ray on the back (dorsal) fin of the viperfish is very long and flexible. It has a blob-like tip that glows in the darkness. Small creatures come to investigate and the viperfish grabs them in its wide, gaping jaws lined with long, needle-shaped teeth. The viperfish looks fearsome and is one of the larger predators of the ocean depths.

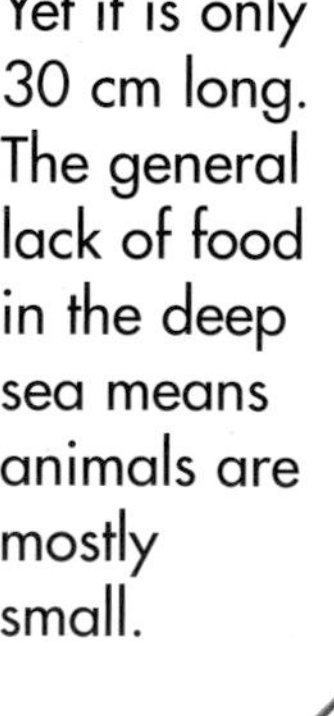

Yet it is only 30 cm long. The general lack of food in the deep sea means animals are mostly small.

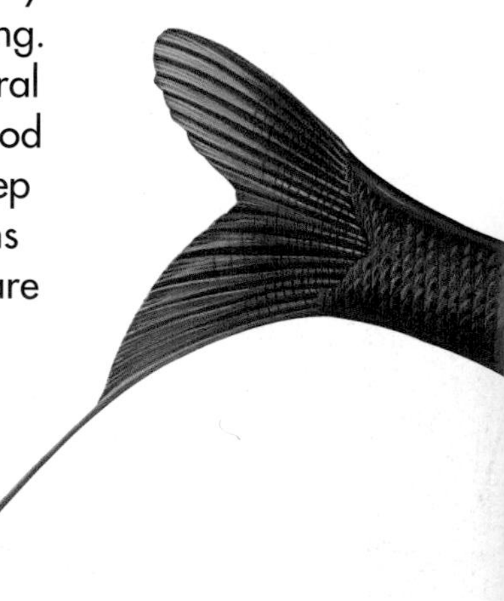

The tripod fish is one of the deepest dwelling of all fish, found more than 6000 m below the surface.

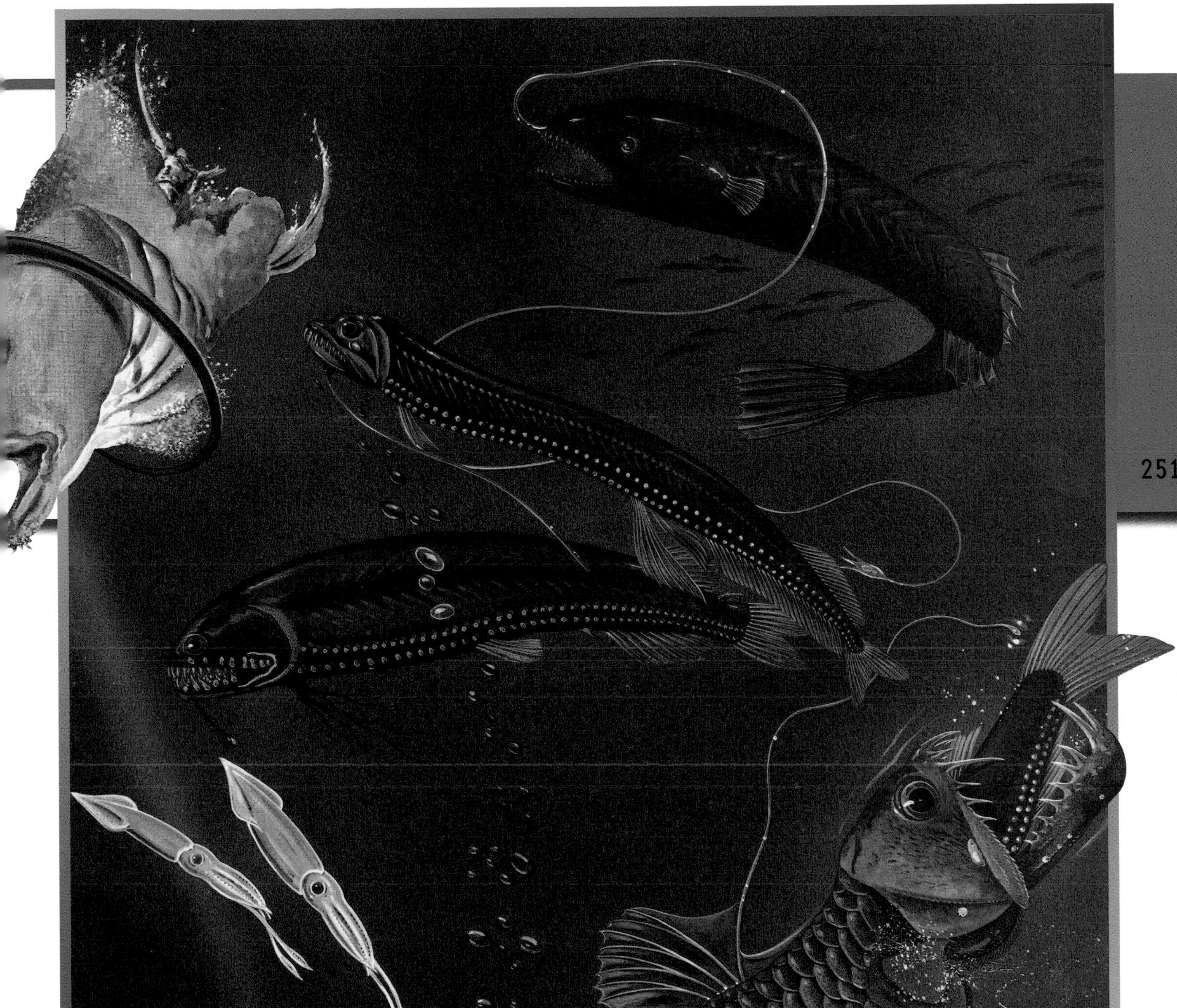

Index

Index

Index

Index

Index

T

U

V

W

Y

Z